GENES, MEDICINE, AND YOU

Alvin and Virginia Silverstein

ENSLOW PUBLISHERS, INC.

Bloy St. & Ramsey Ave.
Box 777
Hillside, N. J. 07205
U.S.A.

P.O. Box 38
Aldershot
Hants GU12 6BP
U.K.

Library of Congress Cataloging-in-Publication Data

Silverstein, Alvin.
Genes, medicine, and you.

Bibliography: p.
Includes index.
Summary: Describes how advances in genetics are being used in screening and counseling and discusses the medical and ethical aspects of genetic engineering.
1. Medical genetics--Juvenile literature.
[1. Medical genetics. 2. Genetic engineering]
I. Silverstein, Virgina B. II. Title.
RB155.S48 1989 616'.042 88-37353
ISBN 0-89490-154-0

Printed in the United States of America

10 9 8 7 6 5 4 3

Illustration Credits:
Agricultural Research Service, USDA: pp. 107, 146; Division of Medical Genetics, University of Medicine and Dentistry of New Jersey--Robert Wood Johnson Medical School: pp. 12, 20, 29, 83; Courtesy of Genex Corporation: p. 85; Dr. Donald Helinski, University of California, San Diego: p. 149; Courtesy of Susan Jones, The Children's Hospital of Buffalo: p. 98; Medical Center of Vermont: p. 75; National Down Syndrome Society: p. 59; National Institute of General Medical Sciences: pp. 23, 39, 92; University of Nebraska--Lincoln: pp. 52, 120; Drs. Zena Werb and Reiko Takemura, University of California, San Francisco: p. 135.

For Gene Petrosky, Jr.

ACKNOWLEDGMENT

The authors would like to thank Joseph D. McInerney, Director of the BSCS program at The Colorado College, for his careful reading of the manuscript and his many helpful comments and suggestions.

CONTENTS

1

Genetic Challenge

Were you a "perfect baby"? Probably your parents worried a bit, now and then, before you were born. Were you developing normally, or might something go wrong? Chances are that when you finally made your entrance into the world, your parents could breathe a sigh of relief: you had the right number of fingers and toes; your eyes and nose and mouth were all in the right places; you were a breathing, crying little human being.

For most parents, a normal baby is the happy result of pregnancy, and the long months of waiting and worrying are quickly forgotten. For about three in a hundred, though, birth may bring a host of problems. These babies are born with birth defects, some of which are hereditary.

About thirty-five hundred genetic disorders are now known—conditions that can be passed on from one generation to another, inherited by children from their parents. Some of them are obvious at birth, such as a jaw or skull that did not form properly. Disorders of the body chemistry are more subtle and may not be discovered until the child becomes ill or does not learn or grow as quickly as he or she should. Some hereditary disorders are like time bombs, lying undiscovered for years until they suddenly burst out—perhaps at adolescence or even in middle age.

Although most genetic disorders are rather rare, researchers have discovered that heredity also plays a role in our susceptibility to many common diseases. These include infectious diseases, such as tuberculosis, and even the major killers, cancer and heart disease. Genetic factors also have been implicated in some forms of mental illness, as well as in drug abuse and alcoholism. Therefore, it's not surprising that major efforts in medical research have been focused on genetics—learning about how the instructions for forming and operating a complex human being are coded and stored inside tiny chemical structures so small that they cannot be seen without a powerful microscope. Researchers have been learning about genes, these tiny units of heredity, and the controls that turn them on and off according to the body's needs. They have made giant strides in recent decades. Geneticists, scientists who study heredity, can now identify many of the genes associated with particular traits—from the color of a person's eyes to a tendency for blood cells to collapse abruptly into a sickle-shaped form. Progress in identifying genes linked with inherited disorders has been so rapid in the past few years that one researcher was prompted to joke about the "gene of the week." Artificial genes have been made in the laboratory, and genes have been transferred from one organism to another.

New knowledge and skills have made it possible to test for genetic disorders in newborn babies or in a fetus developing in its mother's womb. For some genetic conditions there are tests to identify carriers of the trait—people who may show no signs of the disorder themselves but might have children at risk. In a few cases, knowledge of the chemical basis of genetic disorders has suggested treatments to permit people born with them to live a relatively normal life. Babies born with a condition called PKU, for example, used to be doomed to become mentally retarded. But simple and relatively inexpensive tests, which can be run routinely after birth, can detect PKU, and then a special diet can prevent the buildup of poisons that could damage the baby's developing brain. In the future—perhaps not too far down the road—scientists hope to be able to go further. They are boldly planning ways to repair or replace defective genes—to tackle the causes of genetic disorders rather than just treat their symptoms.

The growing capabilities of genetic medicine have brought with them new responsibilities, social problems, and ethical dilemmas. Couples planning to marry or preparing to have a child may face agonizing choices if they find they are carrying potentially dangerous genes. Screening tests for genetic disorders can have enormous benefits, but they can also be misused to discriminate against groups of people. Doctors and researchers, as well as the community at large, must balance tricky equations of benefit and risk in their efforts to determine whether promising new techniques are ready for use on human patients.

In the chapters that follow, we will explore the nature of heredity, the errors that can occur as the instructions for life are transmitted from one generation to the next, the exciting progress being made in gene medicine for detecting and treating genetic disorders, and the future prospects for correcting these errors by gene therapy.

2

Genes, Chromosomes, and Heredity

While turning the pages of a family album, you may notice some familiar patterns repeating themselves, with minor variations. Perhaps you'll see Grandpa's distinctive nose popping up in half a dozen relatives. Or maybe you'll discover that Uncle John had a funny little cowlick in his hair that looks just like the one you see in the mirror every day.

People have known for a long time that a variety of traits are passed from one generation to another by heredity. Some seem to be inherited consistently, passing from parents to their children in a line that can be clearly followed. Other traits may seem to disappear and then pop up in a grandchild or great-grandchild. A pair of brown-haired, brown-eyed parents, for example, may have a blond, blue-eyed baby. In a case like that, probably there were blonds and blue-eyed relatives on both sides of the family, a generation or two back.

For some traits, such as the color of eyes, the curliness of hair, or a particular shape of the nose or lips, general patterns of inheritance can be worked out by close observation of human families over a series of generations. But for others, such as skin color or height, the patterns are not so clear-cut. And what about less tangible characteristics, such as intelligence or insanity or criminal behavior? Sometimes they seem to run in families, but are they really inherited or just products of the environment? Perhaps the children of smart people tend to be more intelligent than the average simply because they are given the advantage of more intellectual stimulation at an early age. A tendency to withdraw from life into mental illness or alcohol abuse might be learned rather than inherited; criminal behavior is certainly at least partly learned.

Human *pedigrees*—family charts showing the relationships of parents and children over a series of generations—can be helpful in working out the nature of particular traits and the way they are passed on, but in many cases their value is limited. Often human families simply aren't large enough to give reliable patterns, and marriages cannot be arranged for the convenience of testing a scientist's theories of heredity. The first real breakthroughs in working out the laws of heredity came from work with a much simpler organism, the garden pea.

Mendel's Laws of Heredity

A nineteenth-century Austrian monk named Gregor Mendel worked out the basic laws of heredity by growing thousands of pea plants over a period of ten years and observing variations in traits, like the color of flowers, pods, and seeds and the height of the plants. He found that certain traits—such as the color of flowers or the height of plants, for example—seemed to be passed from parents to their offspring as though they were being carried by hereditary units, which could be followed on the plants' pedigrees. Each of Mendel's pea plants inherited two of these units for a particular trait, one from its mother and one from its father.

Mendel called his hereditary units "factors," but later (around the turn of the century) the term *gene* was coined for the unit of heredity, and it was discovered that the genes are found on *chromosomes.* These are small rod-like structures that can be seen inside the nucleus at certain times in a cell's life cycle when it is stained with a special dye and examined under a microscope. When the chromosomes are sorted out by size and shape, there are generally the two of each kind in a cell's complete chromosome set. When the *gametes* (the special sex cells that can combine to produce an offspring) are formed, each gamete receives only half of the complete chromosome set—one of each pair.

When Mendel crossed plants of two pure lines—for example, one from a line that always bore yellow seeds and one that always produced green seeds—he found that one variation of the trait seemed to disappear in the crossbred off-

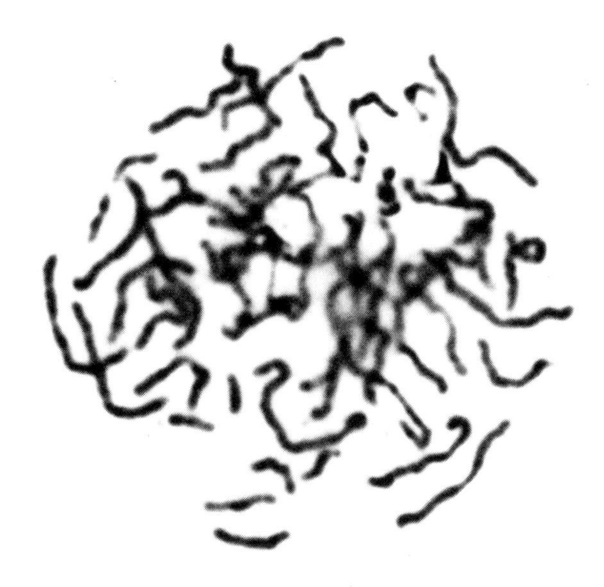

A highly magnified view of a human cell, specially stained to show the chromosomes.

spring. All the seeds obtained by crossing the yellow with the green, for example, were yellow. When plants grown from those yellow seeds were self-pollinated, most of the seeds in the next generation were yellow, but some were green. When all the seeds were tallied, generally two variations were found to occur in almost perfect whole-number ratios—in this case, three yellow to one green. Similar results were obtained for other traits, such as flower color, wrinkling of seeds, and height of plants.

Mendel explained his results by theorizing that in any pair of alternative traits, such as yellow and green seeds, one is *dominant:* the factor (gene) determining it is expressed in the appearance of the trait, regardless of whether the other factor is also present. The variation of the trait that does not appear in the crossbred offspring is called *recessive*. It may show up in the next generation if some of the offspring inherit the gene for it from each parent. (Alternative genes for a particular trait are now referred to as *alleles*. Thus, the genes for the yellow and green seeds in the garden pea are alleles.)

Let's call the dominant yellow-seed allele Y and the recessive green-seed allele g. Mendel's original pure lines would thus be YY (all yellow seeds) and gg (all green seeds). When those pure lines were crossed, each seed received a Y factor from one parent and a g factor from the other. Their genetic makeup—their *genotype*—was thus Yg. All the seeds were yellow because yellow seed is a dominant trait and green is recessive. Scientists refer to this outward appearance as the *phenotype*. It gives some information about heredity, but the information is not complete. A YY seed and a Yg seed would both have the same yellow phenotype. Only a series of matings could give the rest of the information about their genotype.

When both alleles for a particular trait are the same, the genotype is referred to as *homozygous.* This term comes from the word "zygote," the cell that is formed when the male and female gametes join. (The prefix "homo-" means the same, or similar.) When two different alleles are present, the genotype is called *heterozygous.* ("Hetero-" means different.) Thus, both the YY and the gg genotypes are homozygous, and Yg is heterozygous. A yellow pea seed might be either homozygous (YY) or heterozygous (Yg), but a green seed can only be homozygous (gg) because a recessive trait is not expressed unless the allele for it has been inherited from both parents.

The results of matings, just like the flip of a coin, are determined by the laws of statistics. Any particular offspring has a fifty-fifty chance of receiving one particular gene—the Y gene in a pea seed, for example, or a gene for brown pigment

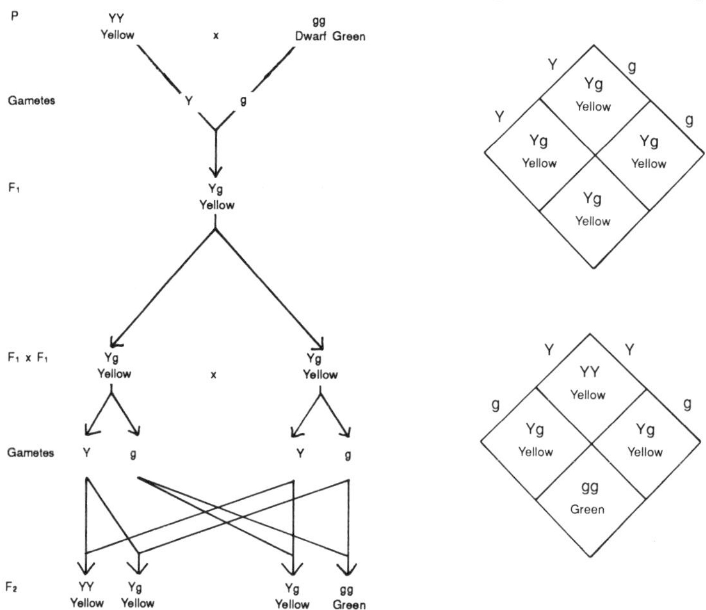

One of Mendel's crosses. P is the parent generation of purebred yellow and green peas; F_1 and F_2 are the first and second generations of offspring. Note that when the heterozygous yellow peas of the first generation are crossed, the recessive trait reappears in their offspring. The box on the right, called a Punnett square, shows a handy way of predicting the genotype ratios of offspring.

in a human eye. Mendel's experiments with peas permitted him to deduce laws of heredity and to figure out genotypes by following the phenotypes over a series of generations because he was dealing with such large numbers. If the number of matings is large enough, the results will tend to even themselves out. Mendel, working with thousands of peas, got almost perfect whole-number ratios. Human pedigrees may not be as informative. Human families are not large enough for the statistical laws to apply, and it is rare for traits to assort themselves in perfect "Mendelian ratios."

The importance of Mendel's work was not recognized at first, and he never knew he would one day be called the "father of heredity." In 1900, long after Mendel's death and close to forty years after he carried out his experiments, three researchers independently rediscovered the articles he had published and brought them to the attention of the world. Then progress in genetics was rapid. Researchers found more convenient organisms to work with than garden peas, such as the tiny fruit fly known as *Drosophila melanogaster* and a red mold called *Neurospora crassa.* These organisms could be raised in bottles or dishes in a modest amount of laboratory space, and they reproduced so quickly that a geneticist could observe many generations in a short time. Much later—not until the early 1940s—it was discovered that a substance called *DNA (deoxyribonucleic acid),* a major component of the chromosomes, carries the hereditary information.

Meanwhile, as experiments in genetics continued, researchers found that although Mendel's laws of heredity apply to a wide variety of organisms from bread molds up to human beings, thcy arc a bit oversimplified.

In each of the pairs of traits that Mendel studied in peas, one of the traits was clearly dominant over the other. Crossing a plant with red flowers with a plant with white flowers always produced red-flowered offspring in the first generation, while the offspring of the next generation, obtained from the crossbred or *hybrid* plants, had a 3:1 ratio of red and white flowers. Yet in a plant called the four o'clock, if red-flowered plants are crossed with white-flowered plants, all the offspring have pink flowers. Taking the cross a generation further shows that the original red and white alleles have not disappeared or changed: The offspring of the pink hybrids are a mixture of plants with red, pink, and white flowers, in a ratio of 1:2:1. In this case a sort of *incomplete dominance* seems to be operating—when the genotype is mixed, both alleles are expressed in the phenotype. There are numerous examples of incomplete dominance in human heredity, such as curly and straight hair: if both curly and straight alleles are present, the person has wavy hair.

The ABO blood group substances are a special case. The A and B substances are both dominant; a person who inherits an A allele from one parent and a B allele from the other will produce both substances and have the blood type AB. But there is still a third allele for this particular trait, the recessive O. This variation of the gene specifies that neither the A nor the B substance will be produced. The homozygous OO genotype is expressed as type O. But there are also two heterozygous phenotypes, when O occurs together with one of the other alleles: AO and BO. In each case, there is one gene producing a blood-type substance (A or B). So the AO genotype is expressed as the blood type A—indistinguishable from the phenotype of someone with the homozygous AA genotype. Likewise, both BB and BO people have blood type B.

Another of the laws Mendel deduced from his studies of peas proved to be rather limited in its applicability. Each of the seven traits that he studied was inherited independently of all the others. If a pea plant with yellow round seeds was crossed with a plant with green wrinkled seeds, for example,

have a normal allele on the other X chromosome to mask the effects of the recessive gene.

The study of linked traits is one of the ways that genetics researchers have been able to work out maps of the genes on chromosomes. The more often two traits are inherited together, the more likely they are to be close together on the chromosome. Here too, however, there are some complications that Gregor Mendel never dreamed of.

Normally, when a cell is dividing in the process that forms gametes, the chromosomes of each pair (referred to as *homologous* chromosomes and representing one inherited from the mother and one from the father) first come together and then separate, to be distributed randomly to the two new cells that are formed. But sometimes, in the process of pairing and separating, the homologous chromosomes get tangled up, and when they untangle, they may switch portions with each other. Some of the genes from the maternal chromosome in the pair may be transferred to the paternal chromosome, while the corresponding genes from that chromosome are attached to the maternal chromosome. This kind of genetic accident, called *crossing-over,* makes an enormous contribution to genetic variation by continually providing new combinations of traits. Crossing-over also provides some valuable clues for genetics researchers working out gene maps. By observing how often genes that normally are inherited together separate, they can deduce how close together the linked genes are on the chromosome. (The farther away two genes are, the more likely they are to be separated by crossing-over.)

Mendel's laws do work fairly well in determining the inheritance of sex. In order to follow that process, let's first take a closer look at the cell's life cycle and what happens to the chromosomes during various phases of it.

Growth and repair of body tissues require cells to multiply. First they duplicate all their important internal structures and then they divide into two new cells, each of which receives one complete set of everything—including the chromosomes that contain the hereditary instructions for the cell and the body as a whole. While a baby is being formed and a child is growing, cell division goes on actively all through the body, but gradually some types of cells stop dividing. In an adult, normally no new nerve or muscle cells are formed, but there is still an active turnover of some other types of cells such as skin and liver cells. This kind of cell division, in which one cell produces two new cells just like it, is called *mitosis.*

Before a cell begins to divide, the characteristic set of chromosomes in its nucleus cannot be seen, even with the proper stains and a good microscope. Instead, there is a tangled, threadlike network of genetic material. In it, each chromosome is stretched out into a long thread, containing millions of atoms. (All together, the stretched-out chromosomes in a human cell have a combined length of about

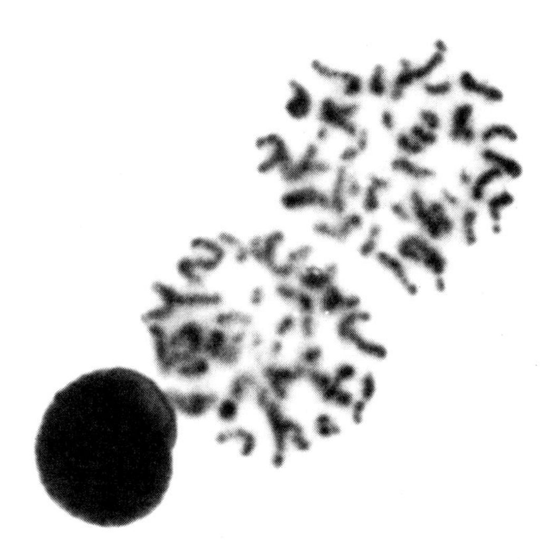

A human cell at the end of mitosis. The chromosomes have been duplicated and separated into two complete sets (disjunction), but the cell has not yet divided. (The cell shown on page 12 was at the beginning of mitosis, when the duplicated chromosomes had just condensed from the thin strands of chromatin.)

174 centimeters or about 68.5 inches—quite impressive when you consider that the individual threads are so thin that you'd need an electron microscope to see them.) As cell division begins, each threadlike strand coils and condenses into the typical rod-shaped chromosome, but now it can be seen that the strand has been duplicated.

The original chromosome and its new copy are attached at a point called the *centromere.* Its name implies that it is found at the middle of the chromosome, which is the case for some chromosomes. But often the centromere is found closer to one end or the other. Its position is always the same for a particular chromosome, and the shape of these joined duplicate chromosomes is one bit of information that can be used to sort and identify them. The parts of the chromosome that extend from the centromere are referred to as the *arms* of the chromosome; when the centromere has an off-center position, the chromosome has a long arm and a short arm.

As cell division continues, the centromeres divide, and the duplicate chromosomes that had been joined move apart, traveling toward opposite ends of the cell. Then the cell itself divides down the middle to form two new cells. Each of them receives one of each duplicated chromosome pair and thus has a full set of chromosomes—identical to the set in the original cell.

When gametes are formed, *germ cells* (special cells found in the ovaries in a female and in the testes in a male) go through a particular kind of division called *meiosis.* The cell divides twice in succession, but the chromosomes are duplicated only once. In the first division of meiosis, the homologous chromosomes pair up and then move apart, much as the separated duplicate chromosomes move in mitosis. But this time the two new cells do not receive identical copies of the chromosome set. Instead, each one has a duplicated half-set, containing one from each homologous pair. Since the homologous chromosomes move at random, a particular new cell is equally likely to receive the maternal chromosome or the paternal chromosome of a particular pair. Considering the whole human set of forty-six chromosomes, a new cell may receive all twenty-three maternal chromosomes and none of the paternal ones, or all twenty-three paternal chromosomes, or any other combination adding up to twenty-three (such as ten maternal chromosomes and thirteen paternal chromosomes). In another cell with the same forty-six original chromosomes, the separation of the homologous chromosomes might produce a different combination, and thus the new cells could wind up with different assortments of alleles for various traits.

The cells formed in the first divisions of meiosis are not gametes. They actually have a total of forty-six chromosomes, two each of twenty-three kinds, with the duplicates still joined at their centromeres. In the second division the centromeres divide, and the duplicate chromosomes separate and move apart; when the cell divides, each new cell receives just one of each of the twenty-three chromosome pairs. In males the division of meiosis produce four *sperm* cells, each with a set of twenty-three chromosomes, one of each kind. In females there is a variation on the process, because the egg must carry not only the mother's genetic information for the child but also a reserve food supply for the earliest stages of its development. So the first division of meiosis in the female germ cells is unequal. Each new cell receives half of the chromosome set, but one of them receives nearly all the rest of the cell contents, while the other gets little more than the chromosomes themselves. It has become a tiny cell called a *polar body*. In the second division, the polar body divides to form two equal products (both polar bodies), but the other cell divides unevenly again, to produce an egg or *ovum* and another polar body. So meiosis in females produces only one egg, plus three polar bodies. The genetic information in the polar bodies is essentially thrown away, but the egg can be fertilized by a sperm, joining with it to form a new cell called a *zygote* and combining their two half-sets of chromosomes to produce a new full set of forty-six.

Let's take a closer look at what happens to the sex chromosomes during meiosis. In a female, whose germ cells start out with two X chromosomes, their separation and division are very much like those of any of the other chromosome pairs. (The twenty-two pairs of non-sex chromosomes are called *autosomes.*) The egg that results from the second division has only one X chromosome—one from the original homologous pair. Which one it is—maternal or paternal—was determined randomly during separation. In a male the situation is quite different. The male germ cells originally contain one X chromosome and one Y chromosome. In the first divi-

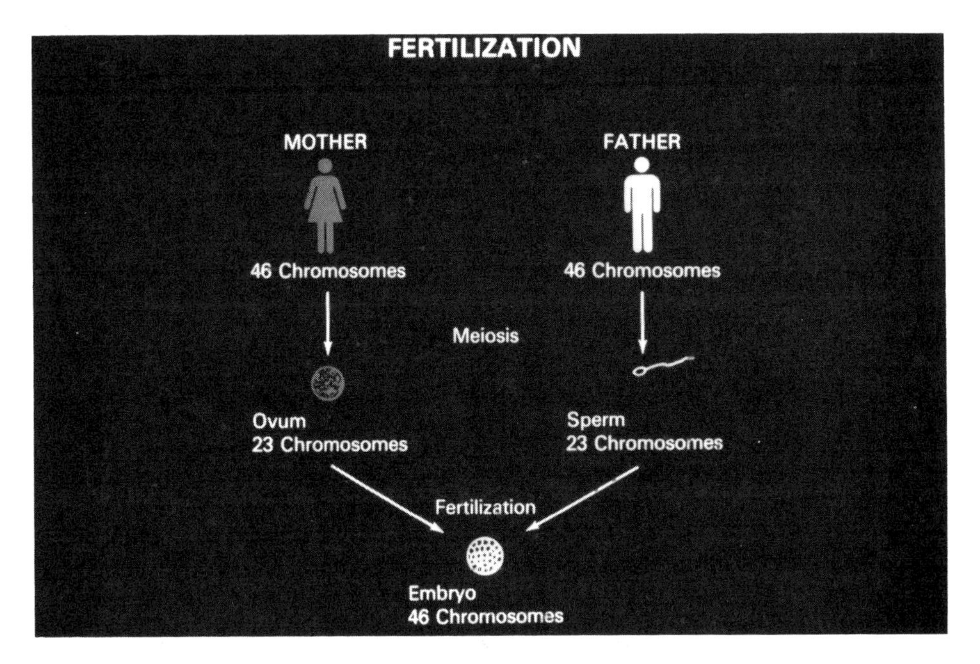

The gametes, sperm and ovum, contain a half-set of chromosomes (23); when they combine, the fertilized egg (zygote) contains the full set of 46 chromosomes.

sion of meiosis, one new cell receives a duplicated X chromosome and the other receives a duplicated Y chromosome. The second division yields a total of two sperm containing an X chromosome and two sperm containing a Y chromosome.

The diagram on this page shows what happens when a sperm and an egg combine to produce a zygote. (The form of the diagram is similar to the ones used earlier in the chapter to illustrate Mendel's laws. This convenient technique for figuring out the results of genetic crosses is called a *Punnett square,* named after the English geneticist who invented it.) There are two possible results of any combination of egg and sperm. The zygote will always receive an X chromosome from the egg (the maternal contribution to the new genotype), but the paternal contribution may be either an X chromosome or a Y chromosome, depending on which the sperm is carrying. The zygote will thus be either XX or XY—either female or male. (Note that it is the father's contribution, the sperm, that determines whether a baby will be a boy or a girl. If a man blames his partner for having a child of the "wrong" sex, he is only displaying his ignorance. He was the one who was responsible for the child's sex.)

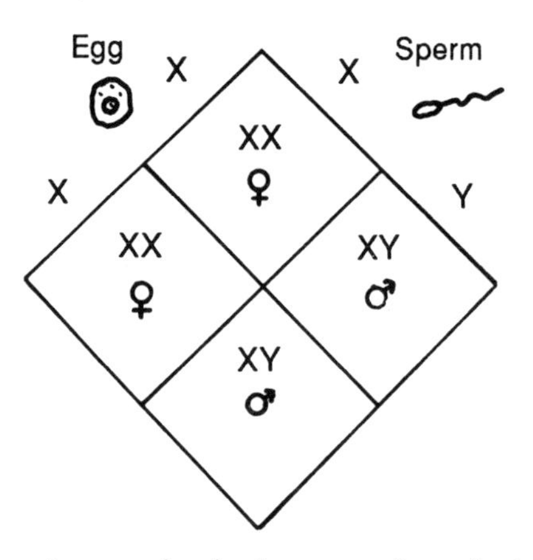

Punnett square showing human sex determination.

The Punnett square predicts that in any particular birth there will be a 50 percent chance of having a boy and a 50 percent chance of having a girl. Those probabilities, of course, are statistical, and they do not work very well on such small samples as the average human family. (Most people know at least one family with all girls or all boys.) In such a large sample as the whole United States population, the ratio should be fairly accurate. Actually, however, it isn't. Birth records show consistently that about 106 males are born for every 100 females, and studies have shown that at *conception* (the fertilization of the egg by a sperm) the ratio is even higher—about 130 males to 100 females. When deviations from the genetic predictions occur, it is usually an indication that some other factors are also influencing the results. In this case, researchers have discovered some possible explanations for the larger number of males. One is that the Y sperm, with their smaller Y chromosome, are a little lighter than X sperm, and they swim faster. In the crowd of sperm that reach the egg first, there will probably be more of the faster-swimming Y sperm, and thus they will have a better chance of fertilizing the egg. There may also be differences in the fragility of the X and Y sperm and their ability to survive under the conditions of the female reproductive tract. At any rate, the differences in numbers are gradually evened out. Males, with only one X chromosome, are more vulnerable to the effects of potentially damaging X-linked recessive traits, and they have a lower survival rate at every stage of development. By the teen years, the beginning of the reproductive age, the ratio of males to females has already fallen to about 1:1.

Multiple Factors in Heredity

Another complication that Mendel did not consider is the fact that some traits are determined not by a single gene but by combinations of genes whose effects supplement and modify one another. It has been estimated, for example, that the inheritance of skin color in humans is governed by at least ten different genes affecting the production and distribution of the dark pigment melanin and resulting in many subtle variations of shades rather than a simple black and white. Eye color, too, is governed by multiple genes. In speaking of multiple genes, we mean separate genes, not an assortment of variations (alleles) at one particular position, or *locus*, on a chromosome. In fact, the genes determining a trait may be scattered over several different chromosomes.

In addition to the effects of multiple genes, traits are also affected by environmental factors. A person may inherit genes that predispose to tallness (tallness is recessive in humans, incidentally, although the inheritance is more complicated than the simple two-allele pattern in Mendel's peas), but a poor diet or illness during the growing years may result in only average height. Many human traits, including a number of inherited disorders, are determined in this *multifactorial* way. Their transmission from one generation to another can be very complicated and may not show a characteristic pattern in pedigrees.

Studies of twins have been helpful in comparing the effects of heredity and environment on human traits. There are two kinds of twins. Identical, or *monozygotic*, twins are formed when a single ovum is fertilized by a single sperm and then, early in the course of development, the dividing cells somehow become separated and produce two babies instead of one. (If more that one separation occurs, identical triplets, quadruplets, quintuplets, or even higher numbers may develop.) Fraternal, or *dizygotic*, twins are the result of fertilization of two ova, each by a different sperm. Monozygotic twins are always the same sex, and in fact any traits that are entirely determined by heredity should be the same in such twins, since their entire chromosome sets are exactly the same. Dizygotic twins have an average of only half their genes in common (the same as any two *sibs*—children of the same parents). About half of the dizygotic twins born are of the same sex and half are of opposite sexes, and though they may share many hereditary traits in common, they may differ in others.

Thus, geneticists who are trying to determine whether a particular trait is hereditary may study the trait in pairs of twins, comparing the *concordance* (a measure of the similarity of given traits of both subjects) in monozygotic and dizygotic twins. For a trait that is purely genetic, the concordance will be complete for monozygotic twins but considerably lower in dizygotic twins. If one twin has a trait determined by a single dominant gene on one of the autosomes, for example, the other twin will be certain to have the trait too if they are monozygotic, but there will be only a 50 percent probability that they will share the same trait if they are dizygotic. For a trait determined by a single autosomal recessive gene, the concordance for a pair of dizygotic twins will be only 25 percent. When traits are governed by multiple genes, the concordance will be even lower. Low values of the concordance for monozygotic twins are an indication that environmental influences play an important role.

Twin studies were one of the bits of evidence that pointed to a genetic cause of Down syndrome (if one twin is affected, a monozygotic twin is always affected but a dizygotic twin is nearly always normal). More recently, twin studies have pointed up the differences between maturity-onset diabetes (highly hereditary) and the juvenile form (influenced by heredity but probably set off by an infectious disease). Twin studies have also helped to cast light on the heredity versus environment controversies about general intelligence and various forms of behavior. In one study, an 87 percent correlation in IQ scores was found for monozygotic twins—the same as for a person taking a repeat test—and only a 62 percent correlation for dizygotic twins, compared with 41 percent for other sibs.

Such studies do have important limitations, though. Twin studies can point to hereditary determination of traits but do not provide any information on what genes are involved, how they act, or how they are transmitted. The results may also be misleading, since twin studies are based on the assumption that monozygotic and dizygotic twins each share the same environment. Actually, however, identical twins often have a special closeness that may not be shared by dizygotic twins, and parents and others may treat them differently. Some twin studies have been designed to minimize environmental influences by comparing monozygotic twins to same-sex dizygotic twins, or to single out the genetic effects by studying twins who have been raised apart.

While some researchers have been trying to deduce general patterns of inheritance from studies of families and population groups, others have been delving into the structures and even the chemicals of heredity—the chromosomes, genes, and DNA molecules. Let's take a closer look at the stuff of heredity.

Molecular Genetics

The photo on the opposite page shows a magnified view of the complete human male chromosome set, sorted into pairs and arranged according to size and shape. This kind of sorted chromosome arrangement is referred to as a *karyotype.* (Notice that in the typical karyotype the chromosomes are all duplicated, with the original and its copy joined at the centromere.) In addition to size and shape, the chromosomes show characteristic patterns of banding in the stained preparations. Different patterns are revealed by various special staining techniques, and they provide information that geneticists can use to identify chromosomes, trace their heredity, and detect some specific defects such as the "fragile X" syndrome, which can cause mental retardation.

The chromosomes and their banding patterns can be seen with an ordinary light microscope. An electron microscope is needed to see the structures inside the chromosomes. Each chromosome has been found to consist of a very long molecule of DNA, in complex with proteins. Inside the chromosome, bits of protein (cores) are linked together like beads

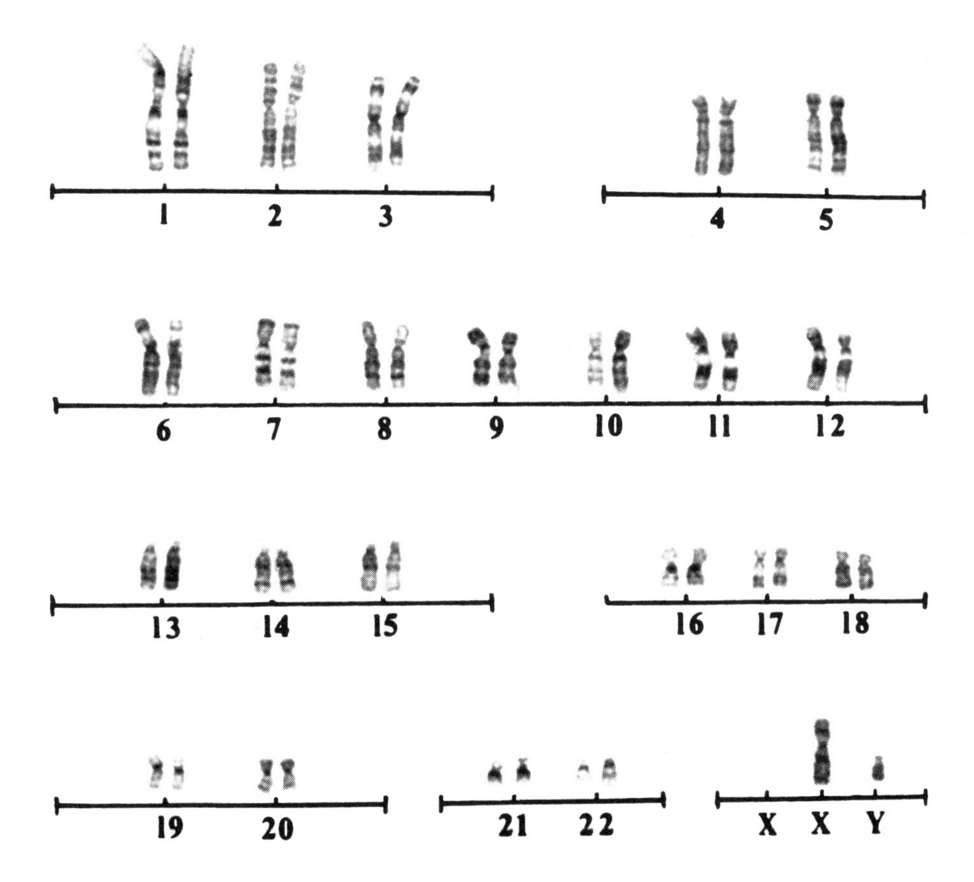

The human male karyotype. This complete set of chromosomes was cut out and sorted from an actual micrograph of a cell. In a female, the last category would contain two matching X chromosomes, instead of the long (X) and short (Y) ones.

on a string, and the threadlike DNA molecule is wound twice around each core. These DNA/protein combinations are intricately coiled, looped, and condensed into thickened structures that show up as bands in stained chromosome preparations.

A DNA molecule is a long chain, composed of units called nucleotides. Each nucleotide consists of three parts: a sugar called deoxyribose, a phosphate group, and a substance called a nitrogen base. Four different nitrogen bases are found in DNA. Two of them, *adenine* (A) and *guanine* (G), belong to the chemical class of *purines;* the other two, *thymine* (T) and *cytosine* (C), are classed as *pyrimidines.* The DNA chain is actually a double chain of nucleotides, curled into a helix and looking rather like a spiral staircase. The "stair treads" are formed by chemical bonds between the nitrogen bases of the two strands, which pair up according to very strict rules: A pairs only with T and C with G. Thus, if the sequence of one strand is known, the sequence of the *complementary* strand bonded to it can be deduced. For example, AATGCAT would be complimentary to TTACGTA. The DNA molecule in each human chromosome is more than 100 million nucleotide pairs long!

When a cell is preparing to divide and the chromosomes are duplicated, a precise copy of each DNA molecule must be made. To do this, the double strand begins to "unzip": The two strands separate, and new complementary strands are built up alongside each one, a nucleotide at a time. Since the units that are added follow the same strict pairing rules, the result is the formation of two DNA molecules, each identical with the original. This copying process is called *DNA replication.*

If DNA contains only four kinds of nucleotides, how can the chromosomes hold all the complicated instructions needed to build and maintain a human being? You might think of these instructions as messages spelled out with an alphabet of four letters, A, C, G, and T. That may seem like a rather limited alphabet, since English uses an alphabet of twenty-six letters. But remember that any word in the English language can also be spelled out in Morse code, using only two letters (dot and dash). A word in Morse code takes up a lot more space than one spelled with the English alphabet, but there is plenty of room in a DNA molecule—each chromosome could potentially spell out a message more than 100 million letters long. The four-letter DNA alphabet allows for a great variety of "words." For example, sixteen different two-letter words can be spelled out: AA, AC, AG, AT, CA, CC, CG, CT, GA, GC, GG, GT, TA, TC, TG, and TT. For a three-letter word there are sixty-four possibilities, and the total goes up exponentially as the word length increases.

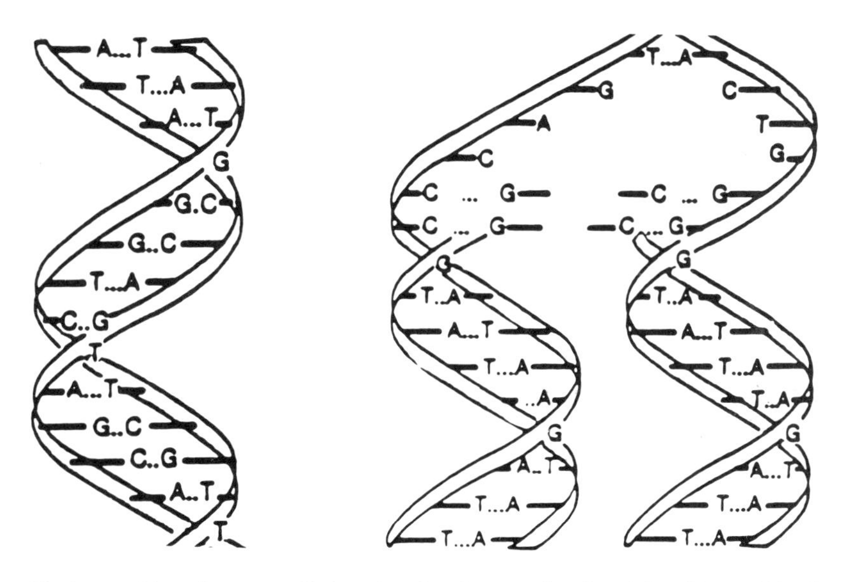

The human blueprints are spelled out in a four-letter code of nucleic acids, combined into long chains and forming the DNA double helix. The pairing of the nucleotides on the two strands follows very precise rules. When DNA is replicated, the two strands unzip, and each is duplicated according to the pairing rules.

Researchers have discovered that DNA messages are spelled out in three-letter units, which are ultimately translated into the amino acid building blocks of proteins. Twenty different amino acids normally occur in human proteins. Since there are sixty-four possible three-letter combinations of DNA bases, several different combinations (called *codons*) can correspond to the same amino acid. Actually, the extra codons are not shared very evenly among the amino acids; two amino acids are specified by only one codon each, and two are each specified by six different codons. Three of the possible codons do not specify any amino acid according to the genetic code that researchers have worked out. Instead, they are read as stop signals, indicating that the growing protein chain is complete.

If you look at the dictionary of the genetic code on page 38, you will notice that the letter U occurs in the codons instead of T. That is because there is another key step in the translation of the DNA message into protein. First the message must be *transcribed* into another type of nucleic acid, called *RNA (ribonucleic acid).* The RNA molecule is also coiled into a helix, but it is single-stranded. It contains the same main parts as DNA: sugar, phosphate, and nitrogen base, but the four kinds of bases in RNA are adenine (A), guanine (G), cytosine (C), and *uracil* (U) instead of A, G, C, and T. When a DNA molecule is being transcribed into RNA, it unzips but only one strand is copied. The bases follow the same pairing rules as in replication, but A pairs with U.

A gene is a portion of DNA that contains the complete instructions for making a protein. Some of the body's proteins are structural, building body tissues or formations such as hair and nails. Others are enzymes, which play roles in body reactions ranging from the digestion of food to the formation and transmission of thoughts. Still other proteins are hormones, chemical messengers that help to control and coordinate the activities of the various body organs and systems. The coding sequences in DNA that spell out the order of the amino acids in a protein chain may be interrupted by noncoding sequences called *introns,* which are edited out of the message at the RNA stage. (The coding sequences are referred to as *exons.* They are rejoined into a continuous sequence after the introns are removed.) In addition, the gene contains flanking regions that regulate the gene, as well as start and stop signals.

Most of the genes in a chromosome are in an inactive, "turned off" state at any particular time. (In fact, some of the genes that are active in early development, forming the basic structures of the body, are then turned off before or not long after birth and never work again.) When a gene is turned on and working, it is first transcribed into a length of RNA. Special *enzymes* (proteins that help chemical reactions to occur) process the RNA, trimming off the introns and converting it to *messenger RNA* (mRNA), which carries the message of the gene out of the nucleus to the cytoplasm, the outer part of the cell. There the messenger RNA attaches itself to structures called *ribosomes,* which contain their own *ribosomal RNA* (rRNA). Matching sequences of codons anchor the mRNA molecule on the ribosome, where it will serve as a pattern, or *template,* for the production of protein. Still another type of RNA, *transfer RNA* (tRNA), picks up amino acids and delivers them to the ribosomes. Each kind of amino acid has its own specific tRNA, which contains a three-letter sequence (an *anticodon*) that is exactly complementary to the codon for that particular amino acid. While the transfer RNA molecules deliver amino acids and hold them in place against the messenger RNA template, the amino acids are linked together to form a protein chain. The gene message has now been delivered.

Chemicals, radiation, and various other influences can act on DNA, producing changes in the bases or causing portions of the chain to be cut out (a *deletion*) or a stray bit of nucleic acid to be added (an *insertion*). Such a change might damage one of the control signals that regulate the gene, turning it on or off inappropriately. A change in a coding sequence might have no effect at all (a change in a base might simply switch the codon to another that specifies the same amino acid), or it might result in a protein so different that the cell—or even the organism—cannot survive. (A deletion or insertion might shift the "reading frame" of the codon sequence. Removal of a letter—the first, for example—from CCT AGA AAC would yield the vastly different message CTA GAA AC, and the protein that resulted would have an amino acid sequence asparagine-leucine instead of glycine-serine-leucine. Changes such as these produce changes in the traits of an organism that are referred to as *mutations.*

Mutations can occur in a body cell (*somatic* mutations) and have very limited effects. But if a mutation occurs in a germ cell, it can affect the development of an organism in the next generation. Occasionally (very rarely) a mutation may result in an improvement that makes an organism better able to survive. This is the basis for evolutionary change through natural selection. Far more often, a mutation produces a negative change, and it may even kill. Most hereditary disorders start out as mutations, and careful studies of pedigrees can sometimes reveal exactly when the mutation occurred.

In addition to changes in the DNA, other things can go wrong in the process of cell division that yields the gametes. A common type is *nondisjunction,* a failure of the homologous chromosomes to separate properly. As a result, one gamete gets an extra chromosome, while another is shortchanged. If that gamete joins with one of the opposite sex to start off a new life, the cells of the new organism may have three of a particular chromosome or only one instead of the normal pair. *Down syndrome,* a condition in which mental retardation is combined with various other changes in the development of the face and body, as well as a shortening of the life expectancy, is caused by a nondisjunction resulting in *trisomy* (three copies) of chromosome 21. Other disorders may result from chromosomes breaking and then rejoining, perhaps with portions transferred to another chromosome *(translocations).* In the following chapter we will survey the varied kinds of hereditary disorders that can result from genetic errors.

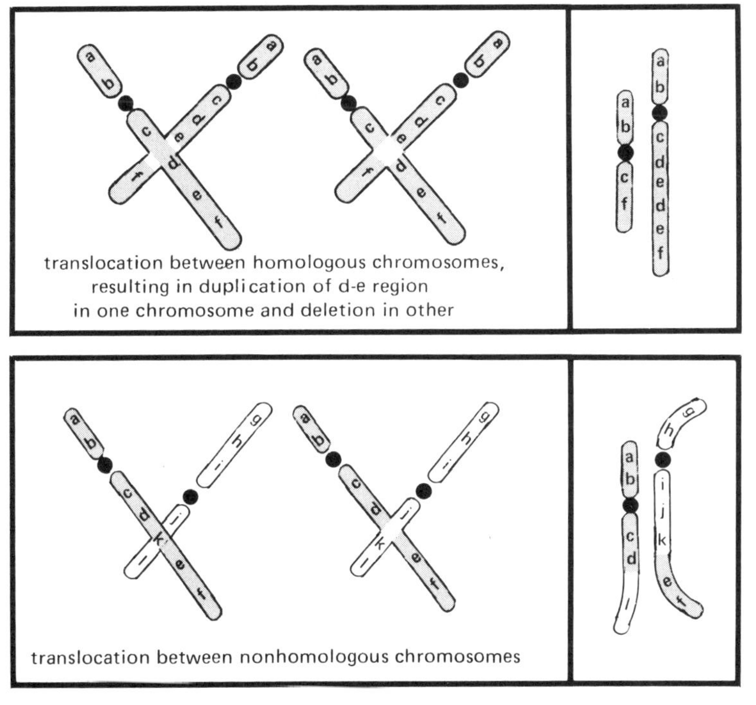

Most cases of Down syndrome are the result of nondisjunction, producing trisomy of chromosome 21. The same condition can also result from translocations, in which the cells receive a portion of chromosome 21 attached to some other chromosome, in addition to the normal 21 pair.

3

Genetic Errors

The DNA in the chromosomes of each of your cells contains a combination blueprint and operating manual for a functioning human being. It is estimated that the human *genome* (the complete set of chromosomes) contains about 100,000 genes, spelled out by several billion base pairs. You might think of it as a 250-page book. How many typographical errors would you expect to find in a book of that size? Each time one of your cells divides, your complete genome—your whole book of instructions—must be copied. The replication process is not a simple Xeroxing, either, but more like resetting every single letter of type by hand.

We have already seen a number of the things that can go wrong in DNA replication and in the division of the chromosomes between the new cells. Considering all these opportunities for error, it is not surprising that researchers estimate each of us carries about five to six potentially damaging genes. What is amazing is that about 97 percent of all babies manage to be born in reasonably normal condition. Most of the potentially harmful mutant genes never produce any noticeable effects because they determine recessive traits, and there is a normal allele on the homologous chromosome to cover for them.

Sickle-cell anemia, for example, is the result of a very small change in the gene coding for *hemoglobin,* the oxygen-carrying pigment in red blood cells that gives blood its deep red color. The form of hemoglobin normally found in adults (Hb A) consists of four long chains of amino acid units, each attached to a complicated iron-containing organic molecule called heme. Two of the hemoglobin chains, the alpha chains, each have 141 amino acid units. The other two, the beta chains, each have 146 amino acid units. In sickle-cell anemia, though, there is a slightly different form of the beta chain: the sixth amino acid from the end is valine, instead of the glutamic acid found in that position on the normal beta chain. If you check the dictionary of the genetic code, you will find that a mutation in a single base in the hemoglobin gene could produce such a change. The messenger RNA codon for glutamic acid is GAA or GAG; valine is coded by GUA, GUC, GUG, or GUU. So a change in the second base of the codon, from A to U, would produce the amino acid replacement.

It seems like such an insignificant change—just one amino acid out of 146 in the beta chains, while both of the alpha chains stay the same. But that substitution produces a change in the electrical charge of that part of the molecule and alters its behavior. When the amount of oxygen in the blood is decreased, sickle-cell hemoglobin (Hb S) forms solid crystals, which distort the shape of the red blood cells. Instead of their usual rounded disk shape, resembling a doughnut without the hole in the middle, they collapse into a sickle shape, like a quarter moon. The sickle cells are fragile and may tear, spilling out their cargo of oxygen-carrying pigment and resulting in anemia. They may also produce sickle-cell crises. Instead of

Codon	Amino acid	Codon	Amino acid	Codon	Amino acid	Codon	Amino acid
UUU	Phenylalanine	UCU	Serine	UAU	Tyrosine	UGU	Cysteine
UUC		UCC		UAC		UGC	
UUA	Leucine	UCA		UAA	*Terminate*	UGA	*Terminate*
UUG		UCG		UAG		UGG	Tryptophan
CUU	Leucine	CCU	Proline	CAU	Histidine	CGU	Arginine
CUC		CCC		CAC		CGC	
CUA		CCA		CAA	Glutamine	CGA	
CUG		CCG		CAG		CGG	
AUU	Isoleucine	ACU	Threonine	AAU	Asparagine	AGU	Serine
AUC		ACC		AAC		AGC	
AUA		ACA		AAA	Lysine	AGA	Arginine
AUG	Methionine	ACG		AAG		AGG	
GUU	Valine	GCU	Alanine	GAU	Aspartic acid	GGU	Glycine
GUC		GCC		GAC		GGC	
GUA		GCA		GAA	Glutamic acid	GGA	
GUG		GCG		GAG		GGG	

Dictionary of the genetic code. Note that more than one triplet codon may specify the same amino acid. (Usually, in such cases, the first two letters of the codon are the same.)

slipping smoothly along in the blood flow as normal red cells do, sickle cells pile up and clog small blood vessels. Tissues downstream of the jam-up, in joints and in the spleen and lungs, are deprived of their usual nourishing blood flow. Painful tissue damage and even death can result from a sickle-cell crisis.

Sickle-cell anemia is inherited recessively. Heterozygotes produce about 45 percent Hb S. The rest of their hemoglobin is the normal Hb A, produced under the direction of the allele on the homologous chromosome. They have enough normal hemoglobin to protect them from anemia and sickle-cell crises, although their blood can be made to sickle under special conditions in a test tube. Such people are said to carry the sickle-cell *trait.* These heterozygous carriers are healthy, but if two carriers have a child, there will be a one in four chance that the child will inherit a sickle-cell gene from each parent and suffer from the disorder. Such sickle-cell homozygotes do not produce any normal hemoglobin.

Kinds of Genetic Disorders

One of the ways that medical geneticists classify hereditary disorders is according to the type of faulty genetic information involved: single-gene disorders, chromosome disorders, and multifactorial disorders.

Single-gene disorders like sickle-cell anemia are caused by mutant genes. They usually show a clear pattern of dominant or recessive inheritance, which can be traced on pedigrees. The faulty genes in dominant disorders may produce changes in structural proteins, such as hemoglobin or collagen (the major protein of the body's connective tissues). Or they may change proteins that play key roles in the body's chemical reactions, such as membrane receptors (which recognize and pick up the messenger chemicals that help to regulate cell and body activities) and enzymes. In typical recessive disorders, on the other hand, the products of the mutant genes are enzymes.

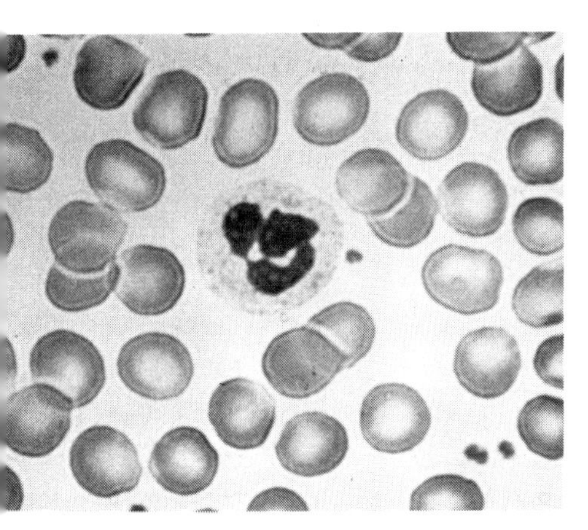
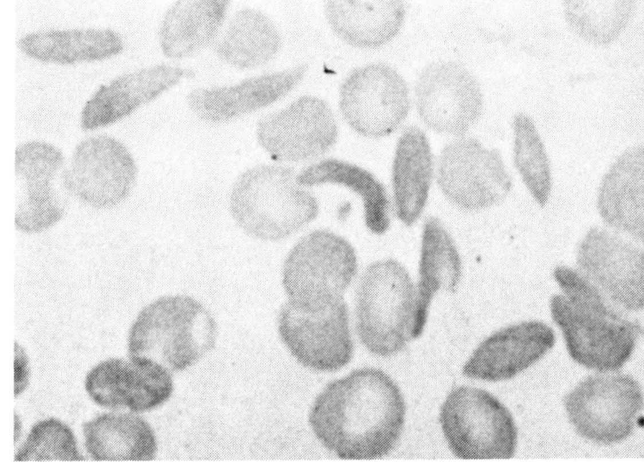

In sickle-cell anemia, some of the red blood cells have a distorted sickle shape. A sample of normal blood cells is shown on the left. (The large cell with a nucleus, in the middle of the micrograph, is a white blood cell.)

Most single-gene disorders are rare, occurring in less than 1 out of 2000 people. However, in certain population groups, such as black Americans, sickle-cell anemia is more common, occurring in 1 out of 500 births. But there is a reason for this statistic. The ancestors of many black Americans came from West Africa, where the mutant gene may be carried by more than 10 percent of the population. In that region malaria is a serious threat, and people with the sickle-cell trait have an advantage. The malaria parasite finds it difficult to live in their blood cells, and the carriers of the trait are thus protected from a dangerous disease. As a result, they have a better chance to survive and pass on their genes, even though some of their descendants may be victims of sickle-cell anemia.

Chromosomal disorders, such as Down syndrome, are the result of abnormal development due to gene imbalance occurring when an extra chromosome or portions of a chromosome are present. Development may be abnormal even though all the genes on the extra chromosome are perfectly normal. Disorders of this kind are rather common, affecting about 7 out of each 1000 live births. They also are the cause of about half of all abortions that occur spontaneously in the first three months of development.

Multifactorial disorders are also rather common and are the cause of many birth defects, such as clubfoot and cleft palate. They may result from the action of multiple genes, or a combination of genetic and environmental influences. These disorders tend to occur in families, but they do not show clear pedigree patterns. In fact, it is sometimes difficult to determine whether they are really genetic or purely environmental. Some environmental factors could also affect a number of family members—for example, the presence of a *teratogen* (something that causes birth defects) in the family's drinking water or in the air of their home.

The various kinds of genetic disorders can also be classified according to their *age of onset,* the time when their symptoms first appear. Some genetic disorders first show their effects during prenatal development. If they are very severe, they may result in spontaneous abortion or miscarriage. Or they may produce a recognizable "birth defect." (A *congenital* disorder is one that developed during the prenatal period and is present at birth. Such conditions may be genetic, or they may be the result of some environmental influence, such as a drug taken by the mother during pregnancy or an infection like rubella.) Some genetic disorders do not appear until the child begins independent life. *Phenylketonuria* (PKU), for example, is a disorder resulting from a genetically determined inability to use one of the amino acids, phenylalanine. Before birth the child receives all its nourishment from its mother's blood. But after birth, when it must eat, digest, and utilize foods, a poisonous phenylalanine product begins to build up in its body.

The genes for still other genetic disorders seem to be turned on at specific ages or stages of development—for example, at puberty or in middle age. Huntington disease, a devastating disorder whose sufferers progressively lose their physical and mental abilities, typically appears in middle age. This is a rare disorder, generally affecting between 1 in 5000 and 1 in 15,000 people. Inherited mutations also contribute to a number of far more common diseases of middle and old age, including the major killers in the United States: heart disease, cancer, and diabetes.

Patterns of Single-Gene Inheritance

Single-gene disorders, both dominant and recessive, typically follow the Mendelian laws of heredity. Strictly speaking, by the way, the terms dominant (expressed in the phenotype whether the genotype is homozygous or heterozygous) and recessive (expressed only when homozygous) refer to traits. But the terms "dominant gene" and "recessive gene" are commonly used for convenience. Disorders due to alleles on autosomes (the non-sex chromosomes) can be traced back through families, and the probability of their occurring in children of a particular couple can be determined in much the same way as the patterns of inheritance in Mendel's peas. But for sex-linked disorders, some special patterns are observed because alleles on the X chromosome do not have homologous alleles on the Y chromosome.

Autosomal dominant disorders are usually caused by rather rare alleles, and it is not very likely that two people carrying the mutation will happen to mate and have children. When one parent is homozygous for the normal allele and the other is heterozygous, carrying the mutant allele and expressing the trait, an average of half the children will inherit the mutant gene and have the disorder, while the other half will be normal. (All the children will inherit a normal allele from the normal parent.) That doesn't mean, of course, that if such a couple has one normal child, the next one will be sure to inherit the disorder (or vice versa). Each new child has the same fifty-fifty chance, just as you have a fifty-fifty chance of turning up "heads" when you flip a coin, no matter how many times you have flipped it before. Probability has no memory.

Examples of autosomal dominant disorders are *dentinogenesis imperfecta,* a condition in which the teeth are brown and wear down readily, and *Huntington disease,* a dominant trait whose carriers may not realize they have it until years after they have given birth to children of their own.

A special case, and a bit of an exception to the general pattern, is *achondroplasia,* a hereditary form of dwarfism in which the head is large with a bulging forehead and a "scooped-out" bridge of the nose and the limbs are short.

Achondroplastic dwarfs have normal intelligence, and marriages between them are fairly common. Adults with this disorder are always heterozygous; the mutant gene produces such severe effects that a double dose of it causes death of the homozygotes in early infancy. A mating of two heterozygotes, Aa and Aa, can result in the combinations: AA + 2Aa + aa. Thus, there is one chance in four that such a couple will have a normal child, one chance in four of having a homozygous achondroplastic baby that will die soon after birth, and two chances in four of having a heterozygous child.

Familial hypercholesterolemia is another example of an autosomal dominant disorder. Heterozygotes have a high level of LDL cholesterol (the "bad" kind of cholesterol that promotes the formation of fatty deposits in the artery walls) in their blood plasma and are at risk for having heart attacks in their thirties or forties. The allele for this disorder is more common than most mutant alleles, occurring in about 1 in 500 people, so it is not too rare for two affected persons to marry. Their children have a one in four chance of being homozygous for hypercholesterolemia and may develop serious coronary artery disease in early childhood. Today, however, they may not be doomed to an early death from heart disease, even if they have inherited the trait. Medical researchers have developed a number of methods for keeping the cholesterol level under control, from special low-fat diets and carefully monitored exercise programs to a number of promising new drugs.

Sometimes an autosomal dominant disorder will appear in a child whose parents show no signs of it. The explanation is that the trait appeared as a result of a new mutation in the germ cells of one of the parents. New mutations are often the explanation for severe dominant disorders. It is estimated, for example, that at least 80 percent of achondroplastic dwarfs are new mutants, born to normal parents. When it can be shown that a new mutation is involved, the risk that the parents will have another child with the same disorder usually is no higher than for the general population. (The risk that an affected person will have children with the disorder is the same as if he or she had inherited the gene from affected parents.)

Autosomal recessive disorders are expressed only in homozygotes, who have inherited a recessive allele from each parent. Such conditions occur most often in the children of two heterozygous carriers of the faulty gene. In such cases the children have a 25 percent chance of suffering from the disorder (homozygous for the trait) and a 75 percent chance of having a normal phenotype. But two-thirds of the apparently normal children (50 percent of the total) will actually be heterozygous carriers and only one-third (25 percent of the total) will be normal homozygotes.

One of the most common examples of autosomal recessive disorders is *cystic fibrosis.* Secretions of various organs and glands of affected people are abnormal, and thick mucus secreted by the bronchi in their lungs makes them very susceptible to pneumonia. In addition, their growth is impaired because of problems with digestive enzymes. Families of cystic fibrosis sufferers must spend hours each day pounding the chests and backs of the patients to break up the heavy mucus. Such techniques, antibiotics to treat the frequent infections, and medications to aid digestion are permitting some people with cystic fibrosis to live a relatively normal life, but many die in their teens or twenties.

The gene for cystic fibrosis is most common among the white (Caucasian) population: The condition affects about 1 child in 2000 births, and about one person in twenty-two in the white population is a heterozygous carrier. (The allele is much less common in blacks and Orientals). When two heterozygous carriers marry, the Mendelian ratios for their potential children are: CC:2Cc:cc, or 75 percent normal (two-thirds of whom are actually heterozygous carriers) and 25 percent affected. Men with cystic fibrosis are usually infertile because of abnormal mucous secretions in the reproductive tract, but some women with cystic fibrosis do have children. In a combination of an affected mother and a normal father, all the children will be normal, with a heterozygous carrier genotype. If an affected woman mates with a carrier of cystic fibrosis, the ratio for their children will be: 2Cc:2cc; that is, half the children will be carriers and half will have the disorder.

Recessive genes that are very rare in the general population may be more common among groups that have been separated from their neighbors for many generations by geographic barriers (the inhabitants of an island, for example) or by religious or ethnic customs. Among American Jews of Eastern European descent, the gene for *Tay-Sachs disease* is much more common than in the general population. This tragic neurological disorder first appears at about the age of six months. The affected child, who was just beginning to babble in response to its parents, sit up, and creep about, gradually loses one mental or physical skill after another, goes blind, and dies within a year or two. The condition results from an inability to produce a single enzyme, "hex A." Without the key enzyme, the cell product on which hex A normally works builds up in the nerve cells of the Tay-Sachs child, interfering with their functioning. Among most Americans, Tay-Sachs disease occurs in 1 out of about 360,000 births, but in Jews of Eastern European origin the frequency is 1 in 3600—a hundred times as common. About 1 in 30 of these Jews is a heterozygous carrier of the allele, compared to only 1 in 300 in the rest of the population. The effects of the gene are not completely recessive; heterozygous carriers produce less of the hex A enzyme than normal people do. But the amount they produce is enough to protect them from the harmful buildup in their cells. The difference in enzyme production is used as the basis of a test for detecting carriers of the gene.

Certain recessive genes tend to occur with a higher-than-average frequency in populations that have been isolated for one reason or another because over the generations marriages have been occurring between men and women who are distant relatives. They share a number of genes in common because of their relationship, and thus their children stand a good chance of doubling up on various genes. This kind of effect is even more pronounced when close relatives mate. Most communities have laws forbidding the marriage of any relatives closer than first or second cousins, but people do not always obey laws, and cases of incest do occur. The children born to very closely related parents—uncle and niece, for example, or brother and sister—are much more likely than usual to have homozygous combinations of genes. (A brother and sister have about half their genes in common, and children of their mating will be homozygous for about one fourth of their pairs of alleles.) Not all children of incestuous matings, however, will suffer from genetic diseases. In ancient Egypt it was the custom for the pharaohs to pass down their rule through a line of brother-sister marriages, and the dynasty survived for thousands of years. But since mutations do occur and many of them tend to have negative effects, the chances are high that incest will lead to genetic problems.

One of the most common of all the known human genetic defects was discovered quite recently. Early in 1988, an international team of researchers reported that they had pinpointed a faulty liver enzyme responsible for bad reactions to a drug used to control high blood pressure. The enzyme normally breaks down drugs into harmless substances that can be eliminated from the body. But about 5 to 10 percent of people of European ancestry are "poor metabolizers." In their bodies, the drug is not broken down effectively and accumulates to poisonous levels. The researchers have identified the gene that carries the blueprints for the enzyme and demonstrated that various changes in it produce at least three different ineffective versions of the enzyme. The "poor metabolizing" trait appears only in people with two of the faulty genes and is thus recessive.

Although the faulty genes are carried by some 35 to 43 percent of the European population, their discovery might seem a rather trivial feat; the blood pressure drug that sparked the bad reaction, in fact, has been withdrawn from the market. But the significance of the genetic defect is far reaching. The "poor metabolizers" also react badly to at least twenty other commonly prescribed drugs. In some cases, as the amount of a drug builds up in the body, its effects are magnified—as though the person had taken a massive overdose. In other cases, when the active form of a drug is actually one of its breakdown products, produced in the body, it may be ineffective even though a seemingly adequate dose was taken.

Moreover, researchers suspect that the metabolizing gene may be a key factor in determining whether a person will be susceptible to developing certain forms of cancer. Lung cancer, for example, can develop under the onslaught of harmful chemicals (carcinogens) in cigarette smoke and pollution. Yet not everyone who smokes develops lung cancer. Paradoxically, the "poor metabolizers" may be protected by their faulty genes from the effects of carcinogens, which are normally broken down in the body into even more damaging substances.

Sex-linked disorders follow some rather special patterns of inheritance, in which the usual distinctions between "dominant" and "recessive" are blurred. Remember that a female has two X chromosomes (XX) but a male has one X chromosome and one Y chromosome (XY). Theoretically, traits could be either X-linked (determined by genes on the X chromosome) or Y-linked (determined by genes on the Y chromosome). Actually, however, geneticists do not yet know very much about genes on the Y chromosome other than those that determine the male sexual characteristics, whereas many genes for nonsexual traits have been located on the X chromosome. So in practice, the "sex-linked traits" are actually X-linked traits. A female, with two X chromosomes, can be either homozygous (both alleles the same) or heterozygous (a different allele on each X chromosome) for a particular X-linked gene. But a male has only one X chromosome and thus can have only one allele for each X-linked gene. Therefore, an X-linked gene will act like a dominant in every male who inherits it, even though it may be recessive in the female carriers.

There is a further complication, too. Among the twenty-two pairs of autosomes (the non-sex chromosomes), all the genes are active, producing their products, even for recessive traits. (Remember that in carriers of the sickle-cell trait, part of the hemoglobin is of the abnormal Hb S variety, and in Tay-Sachs carriers the levels of the enzyme hex A are lower than usual because the mutant gene on one homologous chromosome is not contributing to the enzyme production.) But the situation is different for the pair of X chromosomes in a female. Fairly early in the development of the embryo, one of the X chromosomes in each cell becomes inactive. Which X chromosome remains active and produces its products is apparently random, like the flip of a coin; so on the average, half the cells in a woman's body have the maternal X chromosome working and the other half have a functioning paternal X chromosome. If one of the X chromosomes has a mutant gene, it will produce its changed product (or fail to produce a product) in all the cells that have that X chromosome "turned on." So, depending on what kind of product the gene specifies and what kind of body structure or function it affects, heterozygous females may show a milder form of the disorder or may show no signs of it at all.

If a woman who is a heterozygous *carrier* of an allele for an X-linked recessive disorder mates with a man who does not have that allele, an average of half of her sons will have the disorder and half will be normal. None of her daughters will have the disorder, but about half will be carriers in turn. A man with the disorder cannot pass it on to any of his sons (he gives each of them a Y chromosome, not an X), but about half of his daughters are likely to be carriers. So, like other recessive traits, the X-linked disorder tends to skip generations, but it appears only in sons (unless a carrier woman happens to mate with an affected man; in that case, her daughters have a 50 percent chance of being homozygous for the disorder and suffering from its full effects). Any males in a family group who have the disorder are related to each other only through females.

The most famous pedigree for an X-linked disorder was traced through the royal families of Europe in the nineteenth and twentieth centuries. Queen Victoria of England was apparently a carrier of a mutant gene for *hemophilia.* In this disorder, which occurs in about one in 10,000 male births, the blood fails to clot properly because of a lack of one of the key chemicals in the complicated chain of clotting reactions. Not only is the hemophiliac in danger of bleeding to death after a minor injury, but internal hemorrhages can cause severe ar-

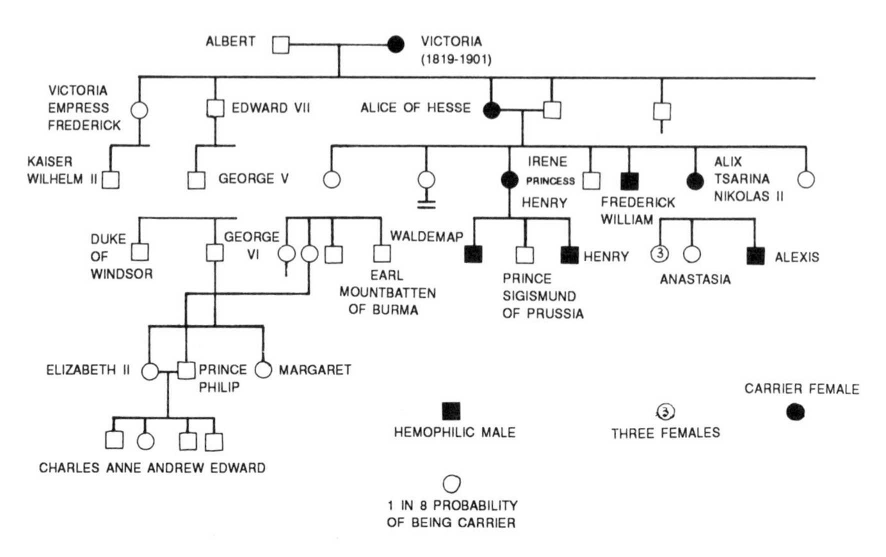

The royal hemophilia pedigree. A mutation in Queen Victoria's germ cells affected world history.

thritis and other painful and disabling symptoms. Seven of Victoria's eight children seemed unaffected, but one of her sons was hemophiliac, and two of her daughters gave birth to sons who also suffered from the disorder. Cases popped up among Victoria's great-grandsons, too. One of them was the Tsarevitch Alexis of Russia. His parents worried constantly about the health of their sickly only son and heir. Desperately seeking a cure, they fell under the influence of Rasputin, who seemed to help the boy but was hated and feared by the Russian people. Eventually the government was overthrown in a revolution, and the little Tsarevitch and his family were murdered by the revolutionaries. An X-linked disorder may thus have indirectly changed the history of the world.

A more common—and even more serious—X-linked recessive disorder is *Duchenne muscular dystrophy,* which affects about 1 in 3000 newborn baby boys. A child with the disorder appears normal at first, but around the time he first starts to walk, he begins to suffer from weakness. This weakness progresses as the muscles waste away. Eventually the child is confined to a wheelchair, and death usually comes before the end of the teens. Women who are carriers of the muscular dystrophy gene rarely show any symptoms. Cases occasionally occur in girls, generally when there has been a new mutation in the father's germ cells. (About one-third of all Duchenne muscular dystrophy cases come from new mutations; the rest show a typical family pattern of X-linked inheritance.)

Studies of six girls with Duchenne muscular dystrophy—the only six girls known to have the condition at the time—led to the discovery of the gene for the disorder, announced late in 1986. Researchers examined chromosome samples from the girls and found that they all had damage in the same area of the X chromosome. The researchers then looked for differences in the sequence of DNA bases in that area, between the X chromosomes of boys with Duchenne muscular dystrophy and those of boys who did not have the disorder. They found a variety of such differences, carefully clipped out the DNA sequences, and used them as *probes* to fish for matching sequences of messenger RNA in fetal muscle tissue. They found one that matched perfectly. Working on the assumption that this mRNA must correspond to a gene coding for an important protein in muscle development, the research team later isolated its product: a protein that is always present in normal muscle tissue but is totally missing in people with Duchenne muscular dystrophy. They named the protein dystrophin.

Medical researchers noticed long ago that there are far more retarded males than females, but it was not realized until fairly recently that many cases of mental retardation are apparently due to an X-linked disorder called *fragile X syndrome,* now thought to occur in as many as 1 in each 1000 male births. Medical researchers examining the chromosomes of two retarded brothers discovered that their X chromosomes had a very unusual and distinctive feature: a pinched-in place on the ends of the long arms, so pronounced that the tips often broke off in the test tube. The same "fragile" X chromosomes were found in two retarded uncles of the boys. Other researchers studying mental retardation began to search for similar forms of X chromosomes in their subjects. The results were very confusing at first—findings that seemed to confirm the pattern could not be repeated when the test was tried again. Eventually it was discovered that the fragile sites appear on the chromosomes only when the blood cells are grown under special culture conditions, and they are observed in only about 35 percent of the cells of a person with

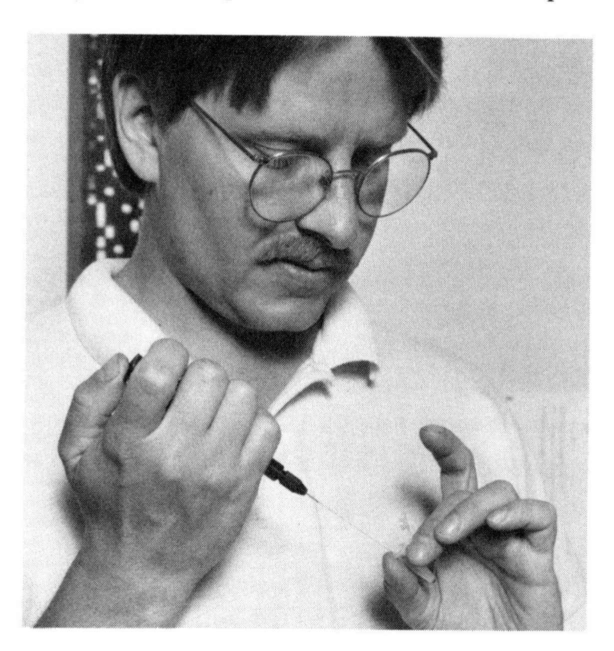

Preparing a sample for analysis of the DNA code.

the disorder. The fragile site can be seen in chromosomes from some carriers of the trait (heterozygous females), but in a smaller proportion of their cells, and some carriers do not show traces of the gene at all.

To further confuse the issue, it was discovered that about 20 percent of men whose X chromosomes have the fragile site are mentally normal, but family histories show that they can transmit the gene to their daughters, who can then give birth to mentally retarded fragile X sons. Most carriers of the gene are mentally normal, but about one-third have learning difficulties. In fragile X children the development of speech is especially delayed, and it has been found that a substantial fraction of the children diagnosed as autistic (with difficulties in communicating and relating to others) have typical fragile X chromosomes.

The fragile X syndrome appears rather often as a new mutation—more frequently than for any other genetic disorder. But these mutations seem to occur only in sperm, never in eggs. Affected males are never new mutants; their mothers are always carriers (in about half the cases as a result of a new mutation) and thus have a substantial chance of giving birth to other affected children.

In addition to mental retardation, the fragile X syndrome produces some typical physical characteristics, including large, protruding ears, a prominent jaw, and large testes after puberty. It is not at all unusual for a single gene to have multiple effects, even though its primary function is simply to direct the production of one protein. Geneticists refer to this production of multiple effects by a single mutant gene or gene pair as *pleiotropy*. In some cases all of the effects can be readily traced to the primary genetic defect.

In phenylketonuria (PKU), for example, a recessive mutation results in a severe enzyme deficiency in homozygotes.

They are unable to produce an enzyme called phenylalanine hydroxylase, which permits people to utilize phenylalanine, one of the amino acid building blocks of proteins. Phenylalanine is a key substance in the body's metabolism. It can be transformed chemically into substances needed for the work of the brain and is also used in the formation of the pigment melanin. In people with PKU these reactions are blocked by the lack of the enzyme. (Heterozygous carriers produce reduced amounts of the enzyme, but enough for these key metabolic reactions to take place.) One secondary result of the enzyme deficiency is a very fair skin and blond hair, due to the lack of melanin production. Another, more serious result is a starvation of the developing brain that leads to mental retardation. As the unprocessed phenylalanine from foods builds up in the tissues and blood, some of it is converted to abnormal products called phenyl ketones, which accumulate and spill over into the urine. They form the basis for the name of the disorder *(phenyl-keton-uria)*. The ketones give the urine a characteristic musty, "mousy" odor. (The disease was first diagnosed in 1934 when a woman in Norway complained to her doctor that her two retarded children always smelled funny, no matter how hard she tried to keep them clean. The peculiar smell was traced to phenyl ketones by chemical tests of the children's diapers.) All these effects are secondary results of the one protein defect.

Another serious hereditary metabolic disorder with multiple effects is *Lesch-Nyhan syndrome,* a rare X-linked condition that occurs in 1 out of 100,000 births. Affected boys lack a single enzyme, hypoxanthine-guanine phosphoribosyltransferase (HPRT), which plays a key role in the synthesis of purines. (Two of the bases in DNA, adenine and guanine, are purines.) The HPRT enzyme helps to regulate purine synthesis; without it, too much purine is formed and builds up in the body. Some of the excess passes out in the urine in the form of uric acid. Some may accumulate in the joints, producing gouty arthritis, or in the kidneys, producing kidney stones. But far more terrible are the effects on the brain. Boys with Lesch-Nyhan syndrome are mentally retarded, suffer from cerebral palsy, and have a compulsion to mutilate themselves by gnawing on their lips and fingers. They must be kept tied down to keep them from hurting themselves.

The gene for Lesch-Nyhan syndrome has been identified and *cloned* (isolated, transferred to bacteria by recombinant DNA techniques that will be discussed in Chapter 6, and reproduced in multiple copies). Heterozygous carriers of the gene can be identified by testing their HPRT level, and similar enzyme tests on cultured cells taken from the fluid surrounding a developing fetus can identify affected babies prenatally.

In another hereditary disorder, *Marfan syndrome,* the typical effects—abnormally long arms and legs and abnormalities of the eyes and cardiovascular system—are all due to a defect in the elastic fibers of the connective tissue.

In addition to the pleiotropic disorders produced by single genes with multiple effects, there are also some genetic disorders in which a number of different genes (not alleles of each other) can produce similar effects. Studies suggest, for example, that there are at least sixteen different types of autosomal recessive deafness, and 10 percent or more of the population are heterozygous carriers of one or more of these genes. In addition, there are some autosomal dominant forms of deafness (including one in which affected people have a distinctive streak of white in the front of their hair), X-linked forms, and some developmental abnormalities that probably are not genetic. Infections such as rubella before birth and middle ear infections in childhood can also cause deafness. Deafness in childhood produces difficulties in communication and tends to isolate the deaf from the hearing population. As a result, marriages within the deaf community are common. Often the children of such marriages have normal hearing—there are so many causes of deafness that it is likely the two parents are carrying different (nonallelic) recessive genes, or one may have a hereditary form of deafness while the other does not. Such cases of *genetic heterogeneity* can produce confusing pedigrees and must be taken into account by genetic counselors advising prospective parents about their risks of having an affected child.

Chromosomal Errors

A new era in medical genetics began in 1959, when a team of French researchers discovered that children with Down syndrome have forty-seven chromosomes in each of their body cells instead of the usual forty-six. Later it was learned that chromosomal abnormalities are much more common than had been thought. They are now believed to account for half of all spontaneous abortions, as well as birth defects and other disorders affecting about 0.7 percent of all live births. More than sixty disorders due to chromosomal abnormalities have already been identified.

Various kinds of chromosomal abnormalities can result in genetic disorders. There may be too many or two few of a particular chromosome; such changes in *chromosome number* may involve one or more of the autosomes, the sex chromosomes, or both. Sometimes not all the body cells are affected, or there may be two or more lines of cells that are abnormal in different ways; these conditions are classified as *mosaicism.* (A mosaic is a picture made up of small tiles of different colors, fitted together). Errors of the chromosome number may arise by nondisjunction, when one or more of the pairs of chromosomes fail to be distributed evenly between the daughter cells while the gametes are being formed. In such cases, sorting the chromosomes reveals that either only one of a homologous pair is present *(monosomy)* or there are three—a homologous pair plus an extra *(trisomy)*.

Sometimes an error of cell division can multiply the whole chromosome set, so that each body cell of the child has three or even four of all twenty-three kinds of chromosomes. This kind of multiplication of the chromosome set is called *polyploidy;* a cell with three of each kind of chromosome is *triploid* and one with four is *tetraploid.* (The normal human chromosome set, with two of each type of autosome and two sex chromosomes, is termed *diploid,* and the twenty-three chromosome set of the gametes is called *haploid.*) In some species, especially in plants, polyploidy is rather common, but in humans it is very rare. Only a few human triploids have been born alive, and tetraploids are seen only in embryos that have been aborted spontaneously at an early stage of development.

In addition to errors of chromosome number, abnormalities of *chromosome structure* can also result in genetic disorders. A portion of a chromosome may be lost, resulting in a *deletion* of some of its genes. Or there may be a *duplication* of a chromosome segment. Generally, deletion errors produce more serious effects than duplications. The *cri du chat syndrome,* for example, is the result of deletion of a portion of chromosome 5 and is characterized by severe underdevelopment of the brain and head, resulting in mental retardation, together with various distinctive and abnormal facial features and a cry of the newborn infant that sounds like the mewing of a cat. In one case a male child with cri du chat syndrome had a sister who was mentally retarded but physically normal and was found to have a chromosome 9 with an extra segment corresponding exactly to the missing genes on her brother's chromosome 5. Their mother was found to be carrying a translocation in which a portion of the short arm of one chromosome 5 had been transferred to the short arm of a chromosome 9. The mother was normal, both physically and mentally, because her chromosome errors were balanced—she had all the necessary genes, even though they were a little peculiarly distributed.

Abnormalities in the children of people with chromosome translocations may be the result of a lack or duplication of part of a key chromosome, producing the effect of monosomy or trisomy as in the case just described. People carrying translocations also have an increased risk of nondisjunction because the affected chromosomes do not match their normal chromosomes and may not pair up and separate properly during meiosis. (When translocations in parents involve chromosome 21, their children may have Down syndrome.)

People with chromosome abnormalities usually have a typical physical appearance, together with a characteristic assortment of mental or physical disorders; often they resemble others with the same karyotype more than they do their own brothers and sisters. In *Down syndrome,* for example, affected people have typical facial features, regardless of their own racial origins. In addition, a baby with Down syndrome usually has poor muscle tone, a protruding tongue, short, broad hands with characteristic patterns of the finger and palm prints, and a short stature. Mental retardation is present, and in about a third of the cases there are abnormalities in the structure of the heart. (Children with Down syndrome are fifteen times more likely to develop leukemia.) Older people with the syndrome typically suffer from premature senility, with the same kind of shrinkage of the brain and neurofibrillary tangles seen in patients with Alzheimer's disease, a loss of mental and physical faculties that typically affects elderly people. People with Down syndrome have a shortened life expectancy, living to an average age of only forty.

The average age of mothers giving birth to a child with

A child with Down syndrome (on the left) with a friend.

Down syndrome is about thirty-four, much higher than the average maternal age in the general population. Nondisjunction seems to occur more often in older women. In women under thirty there is a likelihood of only 1 chance in 750 of having a child with an abnormal chromosome number. But later—especially after thirty-five—the odds rise sharply. In the forty to forty-five age group, the chances of having a baby with an abnormal chromosome number have risen to 1 in 50, and over age forty-five they are 1 in 20. (In each age group, about half the abnormal cases will be children with Down syndrome.) These statistics do not mean that all such children are born to older mothers. Actually, although the risk is much lower for mothers in their twenties, they have many more children than older women do, and thus they give birth to substantial numbers of Down children. For the population in general, the frequency of Down syndrome averages about 1 in 800 births. In the United States and Canada, at least 50 percent of pregnant women over the age of thirty-five are currently choosing to have prenatal tests for chromosome abnormalities.

In addition to the mother's age, heredity may also play a role, since some families have a much higher-than-expected number of trisomies, and a woman who has already given birth to a child with Down syndrome has a higher-than-normal risk of having another. It may be that such women have inherited a defect in the genes that normally control the correct separation of the chromosomes.

Nondisjunction of the *sex chromosomes* does not usually cause as serious developmental problems as when one of the autosomes is involved. Perhaps the reason is that the genes of the Y chromosome determine only sex characteristics, and one of the X chromosomes in each cell of a female is normally "turned off."

In *Turner syndrome* one of the gametes is shortchanged

when the paired chromosomes separate and fails to receive a sex chromosome. Combination with an X-carrying gamete produces an XO genotype (X monosomy). Many XO embryos are spontaneously aborted early in the pregnancy, but some survive. People with Turner syndrome are female, but they do not mature sexually unless they are given doses of female sex hormones in adolescence. (Even with such treatment, an XO woman cannot have children, although the new birth technology options such as in vitro fertilization, discussed in Chapter 4, may make this possible.) People with Turner syndrome have a short stature but can be helped to grow to a normal height with doses of anabolic steroids. They also have a webbed neck and characteristic facial features, but their intelligence is usually normal.

The XXX genotype (trisomy) produces a normal-looking female, who may be taller and thinner than the average and may be somewhat mentally retarded. Some XXX women are infertile, but others have borne children (who usually have a normal karyotype).

Klinefelter syndrome is another trisomy, XXY. This genotype produces tall, rather feminine-looking men whose sex organs do not mature enough for them to father children. Boys with Klinefelter syndrome tend to have a lower than normal verbal IQ and may have difficulties in their schoolwork. They may be less aggressive and active than XY boys. Some variations of Klinefelter genotypes have been observed, including XXYY, XXXY, and XXXXY. Generally, the more X chromosomes, the more abnormal the child's development and the more severe the mental retardation.

There has been a great deal of controversy about the XYY genotype. It was widely discussed in the news when doctors reported (incorrectly, as it turned out later) that the mass murderer Richard Speck had this genotype. Studies have shown that the XYY genotype occurs in inmates of prisons and mental institutions in a far larger percentage than in the general population. Researchers have speculated that the extra Y chromosome makes a man a "supermale," extra aggressive, a real "criminal type." Studies of a small group of XYY boys followed from birth showed that they had normal intelligence but tended to have problems in school; their parents complained of temper tantrums, high activity levels, and moodiness. It is hard to say how many of those complaints were due to the parents' own expectations. Parents who learn that their sons have an XYY genotype are often so worried about the implications that some doctors feel parents should not be told. Many XYY men are perfectly normal, law-abiding people with nothing in their behavior or appearance to distinguish them from XY men.

Chromosome breakage syndromes are a group of autosomal recessive disorders in which the chromosomes have a tendency to break readily in cells being cultured, and there is a defective DNA repair system. In these disorders growth is usually retarded, and there is an increased risk of cancer. In *Fanconi anemia,* for example, retarded and abnormal development is combined with a high level of an immature form of hemoglobin in the blood; affected people have an increased risk of leukemia and lymphomas, cancers of the blood-forming tissues. In *ataxia-telangiectasia* (AT), another disease in this group, there is a progessive damage to the cerebellum, the part of the brain that coordinates movements. Affected people have difficulty controlling their arms and legs. This is a rather rare disease, currently affecting only a few hundred Americans. But the recessive gene responsible for it is found in thousands, and heterozygous carriers have a greatly increased risk of cancer. Recent pedigree studies suggest that the AT gene may be responsible for about 8 percent of all breast cancer, about 10,000 cases in the United States each year.

4

Genetic Factors in Common Diseases and Behavior

"Has anyone in your family had lupus?" Robyn's doctor was trying to make sense of a puzzling set of symptoms, and one possible explanation was systemic lupus, a disease that tends to run in families. Robyn was unable to answer him, though, because she was adopted and knew nothing about her biological parents. More than a year later, after a long and complicated search, she managed to trace her biological mother and was able to find out more about her genetic background. The answer to the question about lupus that had prompted her quest was no—no relative had suffered from that disease. But Robyn learned that her mother had diabetes, obesity, and high blood pressure, and her father had died of a heart attack in his forties. All this was useful information, for with each passing year researchers are discovering more evidence that a person's heredity can play a major role not only in relatively rare genetic disorders but also in common diseases like cancer, heart disease, and diabetes. Mental illness and even certain forms of normal behavior also appear to be at least partly determined by a person's genes, interacting in complex ways with environmental influences.

Genetics of the Killer Diseases

The 1980s have brought a revolution in cancer research, based on the startling hypothesis that healthy humans, as well as other living creatures, carry within their body cells a number of genes potentially capable of causing cancer—*oncogenes.* Normally these genes exist in the body in a harmless form, sometimes "turned off" and sometimes taking part in the carefully controlled processes of normal growth and repair. Dozens of oncogenes have been isolated (including about twenty human oncogenes), and their protein products have been determined. Some of these are enzymes, which cause changes in other protein molecules. Others are DNA-binding proteins that may turn off particular genes. Still others are very similar to growth hormones or the cell surface receptors that interact with them and give chemical instructions that tell the cells to grow. Researchers believe that there are probably between thirty and fifty potential oncogenes in a normal human cell.

A harmless "proto-oncogene" can be activated into a cancer-causing gene in various ways. University of Minnesota researcher Jorge Yunis has found that 70 percent of oncogenes are located near weak points on chromosomes—hereditary regions where the DNA molecule may break or where portions of it may be rearranged into new combinations. Such damage to the chromosome might remove a potential oncogene from the influence of the genetic control mechanisms that normally govern its action. Or a proto-oncogene might be moved next to another gene that has an activating effect. Indeed, many researchers believe that the formation of a cancer cell is a complicated process requiring the cooperation of at least two different oncogenes.

A number of chemical substances, such as the benzpyrenc found in browned meat and in automobile engine exhausts, can produce changes in DNA that might activate an oncogene. (Chemicals that cause cancer are referred to as *carcinogens.*) Radiations can produce similar effects. Viruses may also play an important role in activating oncogenes and even in creating new ones by picking up a bit of the cell's DNA and carrying it off to other hosts. Once activated, some oncogenes give cells the ability to go on growing indefinitely; others affect the way cells interact with their neighbors. Researchers are still working out the details; at any rate, the result is a cell gone wild, multiplying out of control into a tumor mass that piles up untidily and invades neighboring tissues.

The oncogene concept has provided the basis for much more specific knowledge of how heredity and environment interact in the transformation of normal cells into cancer cells. Researchers have discovered, for example, that petroleum-based products, including some commonly used pesticides and insecticides, damage specific sites on chromosomes 5 and 7. Theoretically, the chromosomes of people who come in contact with such products—workers in factories producing them, or farmers who apply them to growing crops—could be tested to see if they have any weak spots on those chromosomes. Those who do would be at increased risk of developing leukemia or other cancers and could be advised to take special precautions or change jobs.

Tobacco smoke produces its damage on a different chromosome, while radiations cause their own specific types of chromosome damage. Genetic tests could thus warn smokers and sunbathers if their heredity had made these activities particularly risky for them.

Meanwhile, cancer researcher Yunis is already using genetic tools to help doctors decide how best to treat their patients. He has discovered characteristic chromosome rearrangements in leukemia and lymphoma cells that signal whether a patient is likely to go into remission (a state in which signs of cancer disappear, at least temporarily) or to die within a few months. In another study, sponsored by the National Cancer Institute, researchers at several medical schools found that children with neuroblastoma who had only one copy of a particular oncogene (N-myc) in their cells had a good chance to survive for at least eighteen months with no progressive growth of their cancers. But chances for survival went down for those whose cancer cells had multiple copies of that oncogene; those with more than ten copies of the N-myc gene had only a 5 percent chance of surviving. With that kind of information, doctors can decide whether to use a very aggressive treatment for cancers likely to grow rapidly or a milder one for patients with only a single copy of the oncogene, who have cancers that grow more slowly.

In 1986 researchers at the Massachusetts Institute of Technology announced the discovery of a gene that *prevents* the development of cancer. In studies of *retinoblastoma,* a cancer of the eye that usually strikes children in their first few years, the researchers discovered that a portion of chromosome 13 was often missing in the karyotypes of children with this disease. A gene normally found in that part of the chromosome works to inhibit cell growth. The normal trait is dominant: Even if the gene is missing on one of the chromosomes 13, the gene on its homologous pair produces enough of the growth-inhibiting product to keep the cells normal. But if an accident happens to the normal chromosome in one of the eye cells—perhaps damage by a chemical or by radiation—then the cell runs wild, multiplying to form a cancerous tumor.

Other research teams are studying another cancer-preventing gene, on chromosome 11. This gene produces a product that prompts cells to *differentiate,* developing into a typical kidney, liver, or muscle cell according to its built-in instructions. (Cancer cells, on the other hand, tend to *de*differentiate, becoming more primitive and generalized.) Sophisticated genetic experiments have revealed a number of other tumor-suppressing genes; researchers now believe that this category may ultimately prove to be even more varied than the cancer-causing oncogenes. In studies of cancer-preventing genes, scientists hope to find ways of turning them on to transform cancer cells back into normal cells.

For many years, polls regularly showed cancer at the top of the list as the disease people feared most. Recently, however, it has been displaced by a newer and even deadlier disease, AIDS. In *acquired immune deficiency syndrome* (AIDS), a virus attacks one of the key parts of the body's system of defenses against invading germs and cancer, a type of white blood cell called helper T cells. In one of the complex interactions of the immune system, helper T cells stimulate a different type of white blood cells (B cells) to produce *antibodies,* proteins that attack an invading microbe or a foreign chemical. But the AIDS virus penetrates the T cells themselves, turning them into tiny virus-producing factories and ultimately destroying them. As increasing numbers of helper T cells are destroyed, the body's immune defenses are weakened, and the person falls ill from a variety of "opportunistic infections"—ones that are normally very rare because a healthy body is able to fight off the microbes that cause them. The AIDS virus can also attack and destroy brain cells.

The AIDS virus is found in blood and various other body fluids, including the semen that is produced in a mature male's reproductive tract and transferred during sexual activity. Thus, AIDS infections can be spread by sexual intercourse and also under conditions when an infected person's blood can enter the body—in blood transfusions or in the sharing of contaminated needles by drug addicts. Fortunately, the AIDS virus is not spread by casual contacts, like shaking hands, working together, or even eating together; and tests for antibodies against the virus are now routinely used to screen the blood for transfusions. Even so, the numbers of new AIDS cases (and deaths) are still rising rapidly, and it is estimated that more than a million Americans have already been infected. (It is not yet known how many of them will actually develop the disease.)

In 1987 a group of British researchers working on the puzzle of why some sex partners of AIDS victims come down with the disease but others do not discovered a striking pattern in the varieties of a particular protein on the surface of blood cells. People can inherit six different combinations of three main subtypes of this Gc protein, produced by a gene on chromosome 4. Among the group in the study, all the sex partners who showed no signs of infection despite long-term intimate contact with AIDS victims had one particular variety of the Gc protein, and none of the AIDS victims studied had

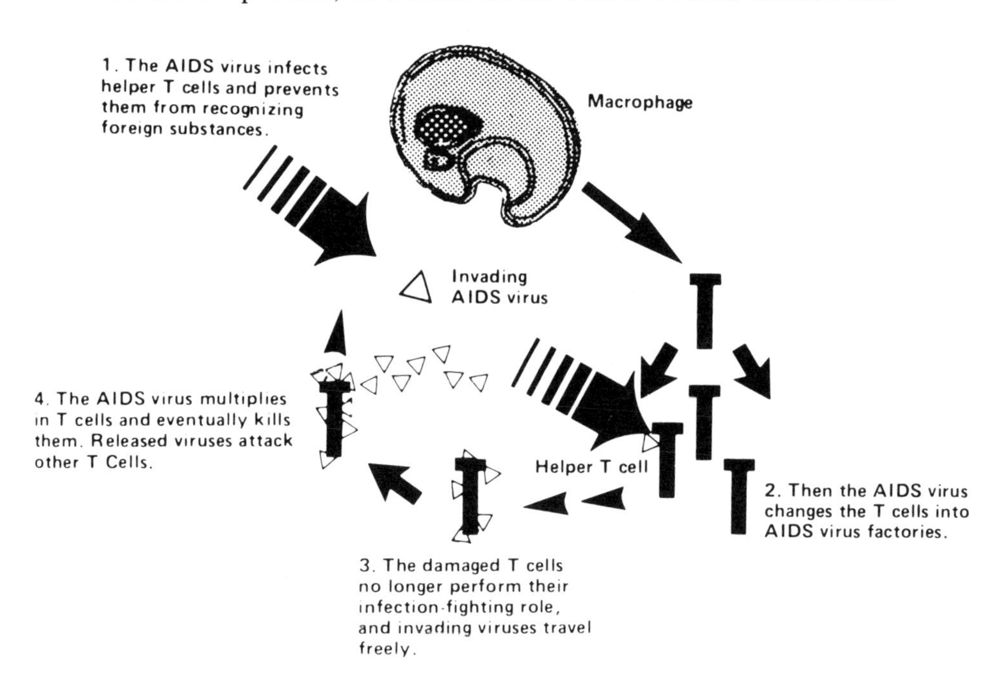

When the AIDS virus invades helper T cells, the normal immune defenses break down.

this variety. Thus, this type of Gc protein seemed to be providing some protection against AIDS. One of the other varieties of the protein seemed to make people more susceptible to developing AIDS and suffering from serious complications. For people with the remaining types, the risk seemed to lie between those two extremes.

This new finding agrees very well with the explosive spread of AIDS in Africa: Blacks from Central Africa are nearly ten times likelier than Caucasians to have the "more susceptible" protein. The researchers caution, however, that even the "more resistant" protein does not provide complete protection from AIDS infection, so a Gc test would just indicate relative risks. But the discovery may also help to reveal more about the nature of the AIDS virus and the way it attacks and destroys cells. The "more susceptible" protein contains a lot of a substance called sialic acid, whereas the "more resistant" protein contains none at all and the other varieties contain intermediate amounts. The researchers believe that sialic acid may help the AIDS virus bind to the helper T cells and invade them.

While cancer and AIDS are the most feared diseases, *heart disease* is actually the number one killer in the United States and the other developed countries of the world. You have probably seen a number of ads claiming that eating "low-cholesterol" foods will help to protect against heart disease. Cholesterol is a fatty substance that can build up in the linings of the arteries, narrowing the channel through which blood flows. The blood supply to vital organs is cut down, and eventually it may be cut off altogether, perhaps by a stray blood clot that acts like a plug. When that happens in blood vessels supplying the heart, a heart attack may result; blockage of blood vessels leading to the brain can cause a stroke. But cholesterol does not entirely deserve its bad reputation. It is actually an important body substance, needed for building nerve cells and some hormones and for various other tasks. If you do not eat as much cholesterol as you need, your body will manufacture it from other substances.

Eventually researchers discovered that cholesterol can be "good" or "bad" depending on substances called lipoproteins that ferry it around in the body. Lipoproteins called LDLs cause cholesterol to be deposited in the artery walls, but another type, called HDLs, actually clean cholesterol out of these deposits. There are various ways to shift the balance of LDLs and HDLs in the body toward healthier levels, by changes in the diet, by exercise, and by the use of certain drugs.

Some people have more need of these LDL-lowering measures because they have inherited an increased susceptibility to heart disease. It has been known for some time that heart disease—especially a tendency to have heart attacks at a relatively young age—runs in families. Researchers have recently found that about a dozen genes can contribute to the development of heart disease. Familial hypercholesterolemia is inherited as a dominant gene that makes people unable to remove LDLs efficiently. The mutant gene, present in about 1 in 500 people, can lead to heart attacks in the thirties. In 1985, University of Texas researchers Michael Brown and Joseph Goldstein won a Nobel Prize in medicine for their landmark studies of mutations of this gene. Another, even more common defective gene, is carried by about 1 in 25 people and interferes with the production of HDLs (the "good" lipoproteins). This gene is associated with heart attacks after the age of forty and is present in nearly a third of all heart disease patients. Several biotechnology companies are now working on tests to identify carriers of hypercholesterolemia genes and estimate the risk of developing heart disease. They have also discovered some genetic markers associated with a decrease in the risk of heart attacks.

Genetic factors have also been found in other common diseases, including diabetes, peptic ulcers, and emphysema. Susceptibility to *autoimmune disorders* such as rheumatoid arthritis and multiple sclerosis, in which the body's immune defenses mistakenly attack some of the body's own tissues, is also genetically determined. These conditions result from errors in the system for distinguishing the surface chemicals characteristic of the body's own cells *(self)* and foreign chemicals (*antigens,* which provoke the production of antibodies). When an organ is transplanted from one person into another, its cells will be attacked as foreign invaders unless its characteristic antigens are a very close match to those of the recipient. Researchers have discovered a whole *major histocompatibility complex* (MHC) that determines which foreign tissues will be attacked (rejected) and which will be accepted by the body. The production of this complex is directed by genes in a particular region of chromosome 6 and includes histocompatibility antigens, components of complement (a substance that takes part in the body's defenses against microbes), and Ir (immune response) genes. It is called the *HLA complex* (an abbreviation for human leukocyte antigen), and five known HLA loci have been found on the short arm of chromosome 6. Each of these gene loci has numerous alleles (at least twenty for HLA-A, more than forty for HLA-B, and eight or more for each of the other three, HLA-C, HLA-D, and HLA-DR). With that many alternative possibilities for each locus, the combinations of all five HLA antigens found in humans are extremely varied.

Different types of autoimmune diseases are associated with particular DR alleles of the HLA complex. Other HLA alleles are linked with increased susceptibility to certain infectious diseases, caused by microbes. Both tuberculosis and leprosy, for example (diseases with very different symptoms but each caused by similar types of bacteria, called mycobacteria), occur most readily in people with the HLA-B5, B12, B14, or B15 antigens. The HLA-B27 allele, on the other hand, brings an increased risk of developing chronic arthritis after an infection with *Salmonella,* a common bacterium in food poisoning. This allele, found in about 8 percent of the population, is present in 90 percent of people with *akylosing spondylitis,* a disease in which the joints become inflamed and eventually stiffened by bony growths. Different human races have different distributions of HLA antigens and show different patterns of susceptibility to disease. In some cases a different allele is associated with a particular disease in different populations, but in other cases the same allele is associated with a disease in all human populations. The list of diseases and disorders for which genetic correlations have been discovered seems to be growing with each passing week.

Genetics and Behavior

Studies of twins recently reported by researchers at the University of Minnesota suggest that people's personality traits, while influenced by their environment and life experiences, are actually predetermined to a considerable degree by their genes. In an eight-year study summarized late in 1986, more than 350 pairs of twins were subjected to extensive tests including analyses of the blood, brain waves, allergies, intelligence, and personality traits. The twins included forty-four pairs of identical twins reared apart from infancy and twenty-one pairs of fraternal twins also separated at an early age. Comparing the similarities of these pairs of twins to those of twins reared together provided a unique opportunity to study the relative effects of heredity and environment.

Identical twins raised apart, in different family environments, were found to be surprisingly similar not only in physical characteristics but also in a number of basic personality traits. Leadership qualities, respect for authority, vulnerability to stress, vivid imagination, feelings of alienation or cheerful well-being, and a tendency to be cautious about taking risks were all revealed to be more than 50 percent determined by heredity according to statistical analyses of the twin studies. Environment played a greater role in determining such traits as aggressiveness, ambition, orderliness, and a need for personal intimacy, although heredity had some influence on these characteristics, too. The researchers stress the need for parents and teachers to accept children's individuality and to try to meet the special needs of their temperaments. If one son is very timid and another is fearless, for example, effective parents would try to provide the timid son with experiences to help him become more comfortable with risk-taking while teaching the fearless son to use some sensible caution.

Considering the pervasive influence of heredity on so many aspects of behavior, it is not surprising that researchers have found evidence suggesting genetic causes for many disorders believed to be largely behavioral rather than physical. Many children who have problems in school and are considered lazy or disruptive, for example, are actually suffering from *dyslexia,* a form of reading difficulty in which the brain seems unable to process sequences of letters or numbers correctly. Dyslexics have difficulty in distinguishing words and tend to transpose letters—confusing "band" and "hand," for example, or reading "was" for "saw." Massachusetts Institute of Technology researchers have also found vision abnormalities in dyslexics: While most people focus on a narrow point directly in front of them while reading and see that center area most clearly, people with dyslexia (estimated at between 5 and 10 percent of the population) find it hard to distinguish the word in their center of vision but can see the words off to the sides much better.

In a study of sixteen families whose members had a history of dyslexia, reported by Herbert Lubs of the University of Miami, a specific defect on chromosome 15 was found in about one-third of the dyslexics studied. Researchers are now searching for other genetic defects that may be linked with some of the other two-thirds of dyslexia cases. (Some have environmental causes, such as head injury or brain disease.) If geneticists can devise tests to predict the majority of dyslexia cases at birth, special teaching techniques could be used to help these children learn to read more effectively. Thomas Kemper of Boston City Hospital, for example, has reported success by training dyslexics to use their sharp peripheral vision in reading.

Alcoholism and *drug addiction* are disorders that have usually been viewed as behavior problems but apparently are also strongly influenced by heredity. Biochemist Lawrence Lumeng of the Indiana University School of Medicine has identified the gene responsible for the flushing, headaches, and nausea experienced by some 30 to 45 percent of people of Asian ancestry after drinking alcohol. This gene directs the production of a faulty liver enzyme that permits acetaldehyde, a toxic breakdown product of alcohol, to accumulate in the body. Although no comparable gene has been found to explain why some people can drink socially while others develop a craving for alcohol (and researchers suspect that there is no simple genetic cause for alcoholism), studies of children of alcoholics have yielded some suggestive findings. Even young boys who have never taken an alcoholic drink show distinctively distorted brain wave patterns, similar to those of their alcoholic fathers. College-age children of alcoholics exhibit a lower sensitivity to alcohol than other college students, becoming less tipsy after several cocktails or remaining better able to keep their balance on a wobbly platform. Such observations suggest inherited differences in the way people's bodies process alcohol.

That such differences are primarily hereditary rather than learned within the family environment is supported by a number of studies of adoptees. When the biological parents were alcoholics, adopted children become addicted to alcohol or drugs in far higher percentages than in the general population. No such correlation is observed with the adoptive parents. Thus, genetic factors seem to be far more important in determining addictive behavior than the learned examples of the family. In one study in Stockholm, Sweden, 18 percent of the adopted sons of alcoholics developed severe drinking problems, regardless of whether their adoptive parents drank. This percentage was six times the rate of alcoholism in the general population. The same Swedish study revealed another pattern of hereditary alcoholism: When one or both of the biological parents had mild to moderate drinking problems, both sons and daughters tended to develop similar moderate drinking problems if they were raised by nondrinking or moderately drinking adoptive parents. But if they grew up in the homes of heavy drinkers, they tended to develop severe drinking problems. In such cases, both heredity and environment seemed to play a strong role.

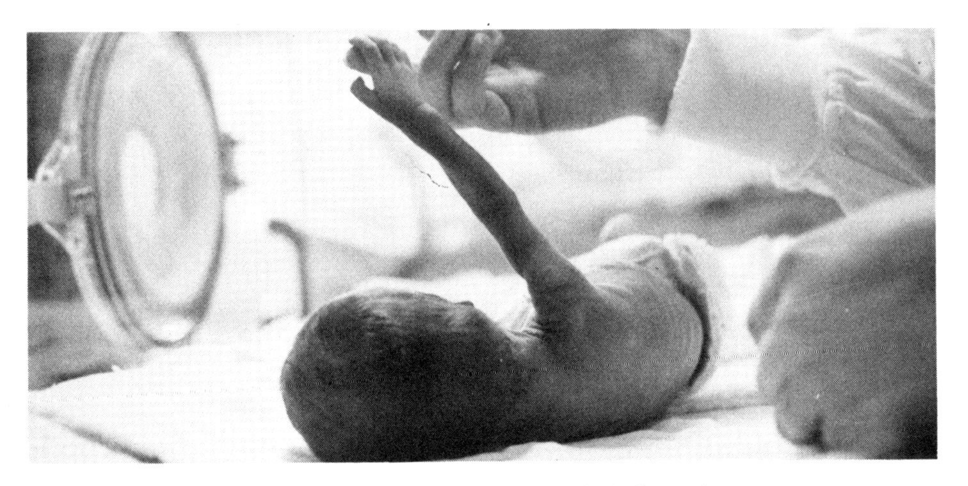

A low-birth-weight baby born to an alcoholic mother.

Researchers caution that heredity can determine only a predisposition to alcoholism; a variety of other factors contribute to determining whether a person actually becomes a problem drinker. (About 60 percent of American alcoholics come from families with no history of alcohol abuse.) Until more precise tests for susceptibility are developed, however, children of alcoholics would be wise to consider themselves at special risk for the disorder and to avoid even social drinking.

The same Stockholm researchers who studied the role of heredity in alcoholism found some further suggestive patterns in the statistical records when they looked for correlations in *criminal behavior* between biological parents and their offspring reared in adoptive homes. Only 3 percent of adopted men born of law-abiding parents and reared in "good" homes had a record of conviction for a crime. But 12 percent of men whose biological parents had criminal records, reared in equally good adoptive homes, went on to commit crimes of their own. For adoptees who had an unfavorable early environment, being shuttled from one institution to another and finally being placed with families of low social status, 7 percent of those whose biological parents were law-abiding committed crimes, while for similar adoptees whose biological parents had criminal records the crime rate was 40 percent.

Such studies are suggestive, but precisely what kinds of genetic factors may be involved is still a matter of speculation. As stated earlier, it has been proposed that sex chromosome abnormalities may be involved—perhaps an XYY genotype leading to an excess production of male hormones and problems of emotional control—but there is little reliable evidence for such speculations so far. In view of the complex behavior involved, such traits as "criminal tendencies" seem likely to be multifactorial.

The genetic evidence is far more definite with regard to some forms of mental illness. In 1987 researchers reported on a decade-long study of a genetically and culturally isolated Amish population in southern Pennsylvania. A defect in the short arm of chromosome 11 was found to be linked with *manic-depressive illness,* also known as *bipolar disorder.* People with this disorder cycle back and forth between periods of deep despair, often suicidal (the depressive phase), and episodes of extreme restlessness, racing thoughts, and delusions of grandeur (the manic phase). The nature of the gene involved has not been determined as yet, but it is inherited as a dominant, and researchers suspect it may be associated with the production or regulation of the enzyme tyrosine hydroxylase. Another report in the same journal showed that the bipolar disorder gene found in the Amish was not associated with the illness in two other populations studied. But several weeks later another research team reported finding a manic-depression gene in five large families in Jerusalem. This gene was located on the X chromosome and may account for one-third of all cases of the disorder. Geneticists now believe that manic depression is not a single condition but rather a whole family of disorders with different genetic causes but similar symptoms.

Meanwhile, National Institute of Mental Health researchers have found that skin cells of manic-depressive patients and their relatives with a history of the condition carry 50 percent more than the normal number of receptors for *acetylcholine,* a chemical that transmits messages between brain cells. The researchers theorize that the patients may also have more acetylcholine receptors on their brain cells, making them hypersensitive to the neurotransmitter and explaining their severe mood disruptions. The study also provided an explanation for the effectiveness of lithium in treating the manic-depressive disorder. When lithium is added to a culture of skin cells from an affected person, it decreases the number of acetylcholine receptors. The NIMH researchers hope to devise a simple skin test to determine whether children of manic-depressive parents will develop the disorder.

One of the most feared conditions associated with growing old is *senility,* a progressive loss of memory, reasoning power, and other abilities. It was thought at one time that senility was part of the natural aging process; yet some people reach their eighties, nineties, and even greater ages with minds still clear and alert. Today it is recognized that some of the physical disorders of old age and the side effects of the drugs used to treat them may lead to memory loss and confusion. And the cases of severe senility often are not natural aging at all, but rather a specific disorder called *Alzheimer's disease.* Examinations of the brains of these people after death reveal some abnormalities: accumulations of deposits (plaques) of an abnormal protein, amyloid; pairs of nerve fibers twisted into unusual neurofibrillary tangles; and deposits of aluminum. There is a loss of brain cells and a decrease in the amount of the neurotransmitter chemical acetylcholine.

The fact that Alzheimer's disease runs in families suggested a genetic involvement, and the resemblance of the neurofibrillary tangles to those observed in Down syndrome called researchers' attention to chromosome 21. In 1987 several independent research teams reported that both the genetic defect in familial Alzheimer's disease and the gene for the abnormal amyloid protein are located on chromosome 21. (The familial form of the disease, which typically strikes people under the age of fifty, is much less common than the form that begins in old age.) Using samples of DNA extracted from white blood cells, still other research teams in the United States and France compared the genetic makeup of normal people, children with Down syndrome, and people with Alzheimer's disease (the more common, nonfamilial form). Both those with Down syndrome and those with Alzheimer's disease had three of the amyloid genes instead of the normal two, although the Alzheimer's patients had an apparently normal karyotype, with only two chromosomes 21. Apparently the amyloid gene was somehow duplicated in these people, producing the same extra dose of hereditary information as trisomy provides in Down syndrome. Researchers do not yet know whether the tripling of the amyloid gene is the cause (or, at least, a contributing cause) of Alzheimer's disease. Probably environmental factors and the body's aging process help to determine at what age the disease develops. Some of the current studies are focused on identifying the factors that modify the timing of the gene's expression. As one Alzheimer's researcher points out, if the expression of the gene could be delayed by just five years, the number of cases of this disorder could be cut in half.

Researchers are also trying to develop tests to assess a person's risk of developing Alzheimer's disease. At Boston University, Miriam Schweber is working on a gene probe test that has been 100 percent accurate in preliminary trials. The probe, only a couple of thousand base pairs long, detects trisomy in a portion of chromosome 21. It gives a positive result for cell samples from people with Down syndrome and Alzheimer's disease and so far has been negative for all other samples tested. Schweber points out that it would be inappropriate to use such a test for prenatal screening, since some people carrying the gene might not develop Alzheimer's disease until the age of ninety. But it can be useful for early detection, when a person is just beginning to exhibit symptoms that might be confused with various other conditions, such as the side effects of blood pressure-regulating drugs or depression.

The genetic tests that are already available and the ethical questions raised by their use will be explored in the next chapter.

5

Genetic Screening and Genetic Counseling

If there were a test that could determine whether you are carrying the genes for a deadly disease that could strike you ten years from now, would you want to take the test? What about a test that could predict your chances of having a child with some disabling or fatal disorder? What would you do if you wanted to have a child and found out you were a carrier of such a genetic disorder? What if, as a couple, you were already expecting a baby and found out that it was likely to be mentally retarded, or crippled with a condition that would require years of expensive and exhausting special care? Would you choose to have an abortion? What if the child were going to be only slightly retarded? Or the wrong sex? What would you do if you were applying for a job and your prospective employer required you to take a battery of genetic tests—with the implication that you wouldn't be hired if your test results showed you were likely to get sick or become disabled? If you were an employer, would you want to hire and train someone who was likely to become disabled after a few years?

When your parents were your age, questions like these would have sounded like good plot ideas for a science fiction story. But now any one of them could be a real dilemma—a choice that you might find facing you one day soon. Genetics has come a long way since Mendel's time. Now there are ways to determine a person's risk of developing many genetic disorders later in life, ways to estimate the likelihood that a couple will have a child with a genetic disorder, and even ways to check on a baby's potential health months before it is born, determining whether it has any of several hundred different inherited disorders. The new testing techniques are part of the fruits of a revolution in biology and medicine. Scientists are now able to probe the human cell down to the molecular level and are taking the first tentative steps toward altering our biochemical blueprints and instruction manuals. This new power can help to ease suffering, but it is also raising some disturbing questions and ethical dilemmas.

Genetic screening is the testing of people without specific symptoms of disease to determine whether they possess a particular genotype. Sometimes the term is also used for specific tests of people at risk for a condition.

Genetic screening may be conducted for research purposes—to test new screening methods or to establish the relationship between a particular genotype and a medical disorder. Genetic screening can also reveal which workers are particularly vulnerable to chemicals or other potentially harmful factors in certain working conditions so that special precautions can be taken to protect them. Genetic tests can pick out people in need of medical intervention. Screening programs may be set up for newborn infants, such as the blood tests for PKU required in some states; with a special diet, such children can be saved from becoming mentally retarded. Screening for genetic susceptibility to cancer or heart disease could result in added years of healthy life for many adults. The main distinction between this type of screening and nongenetic screening (such as routine blood pressure tests) is that the results may be of importance not only to the person tested but to family members as well.

An important application of genetic screening is to determine the risks of genetic disorders in potential children, either through *carrier screening* (testing the prospective parents for the presence of genes associated with hereditary disorders) or *prenatal diagnosis* (the use of special techniques to examine the fetus or cells from it to determine whether abnormalities are present).

Genetic counseling is a process in which parents or relatives at risk for a disorder that may be hereditary are advised about the nature of the disorder and its consequences, their probability of developing and transmitting it, and ways it can be prevented or treated. The counseling process usually includes an analysis of the family history, tests to determine specific risks, advice on the options available, and aid in making decisions about what to do.

Genetic Screening

Since each of us is likely to be carrying a number of potentially damaging genes, there is a possibility that a prospective mate might have a match for one of these "bad" genes. Then, any potential children might happen to inherit a pair of dangerous recessive genes and suffer from a genetic disorder. With some 100,000 genes altogether and the low frequencies in which the damaging mutants occur, the probability of such a genetic mishap is not very great. But it may be multiplied if both members of the couple are of the same racial or ethnic group and thus share more genes in common than the average. Risks also increase when other family members, perhaps in previous generations, are known to have had a particular genetic disorder.

The list of genetic disorders for which tests are available to identify carriers is growing explosively. Some of these tests are simple and readily adaptable to mass-screening programs. The gene for sickle hemoglobin, for example, can be detected by taking a blood sample, placing it under low-oxygen conditions, and examining it under a microscope. If the blood "sickles," then the person is carrying a mutant hemoglobin gene. For Tay-Sachs disease, the levels of the enzyme hex A are determined in skin or blood cells. Heterozygous carriers of the disease produce lower than normal levels of this key enzyme. Similar biochemical tests are available for a number of other metabolic disorders.

Karyotyping, examining the chromosome set from a sample of blood cells, can reveal various kinds of chromosomal disorders. Errors of chromosome number might result in abnormalities in potential children, even in some cases when the person carrying the abnormal karyotype does not show any outward symptoms. Karyotype studies might reveal that a

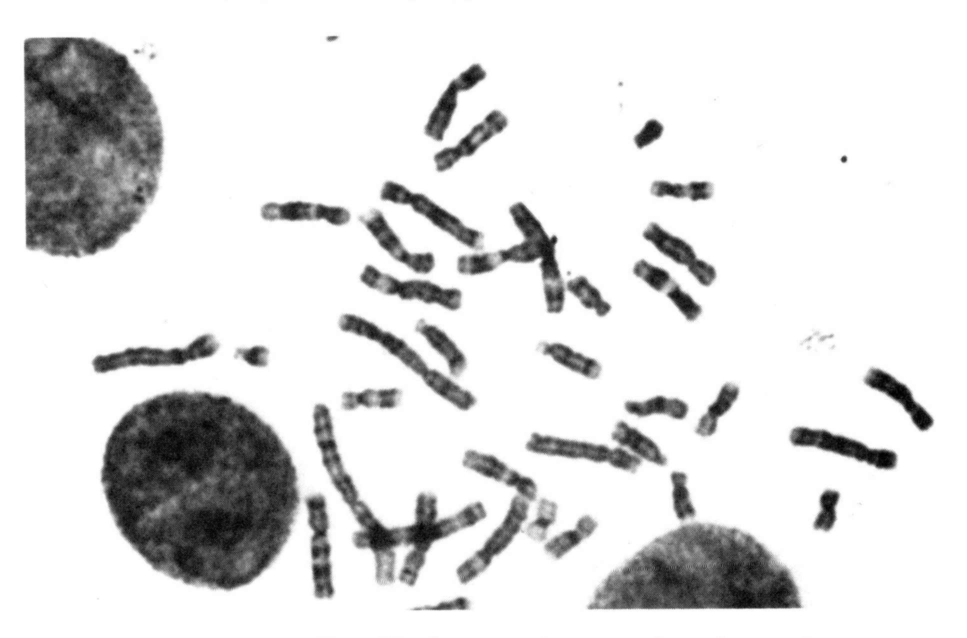

Preparing a karyotype. The cell has been treated to rupture the nuclear membrane so that the stained chromosomes spread out through the cell and can be more readily cut out and sorted. (The large blobs are miscellaneous cell material.)

person is a mosaic, some cells having a normal karyotype and others possessing an abnormal chromosome set. The normal cells may have given the person a normal phenotype, but the mosaic's children might be produced by a gamete with the abnormal set. Or a person might be revealed to be carrying a balanced translocation—say, a portion of chromosome 5 transferred to chromosome 9. When the chromosomes separate during the formation of gametes, the balance might be lost and the child might be lacking genes or have part of a chromosome in excess. The banding patterns that show up when chromosomes are treated with special stains have also been linked with specific disorders, as have other special features like the constrictions near the ends of fragile X chromosomes.

Until recently, these were the main kinds of tests available for genetic screening. This field of medical expertise really began to take off, however, after medical researchers were provided with powerful new tools based on recombinant DNA technology.

Recombinant DNA molecules contain portions of genetic material from different organisms, put together, or "recombined," by a set of special enzymes that scientists use like scissors and glue. The "scissors" enzymes are called *restriction enzymes.* Each type can cut the double-stranded DNA molecule at a particular base sequence. A useful restriction enzyme called Eco RI, for example, isolated from the common intestinal bacterium *E. coli,* makes a cut in the DNA chain between G and A every time the sequence of bases on one of the DNA strands spells GAATTC. Other restriction enzymes have different recognition sites. By using a mixture of restriction enzymes or using several in sequence, scientists can obtain a collection of DNA fragments whose lengths are strictly determined by the "spelling" of the base sequence and the location of recognition sites along it.

In most recombinant DNA work, such DNA fragments are inserted into small circular forms of DNA called *plasmids,* which occur naturally in bacteria. The plasmid is cut open with the same restriction enzymes, so that the cut ends match up with the dangling ends of the cut-out DNA fragment. They are joined together with the other type of enzyme, a *DNA ligase.* Now the DNA fragment is part of a convenient-sized plasmid package that can be inserted into bacteria and reproduced in huge numbers of copies, so that there will be large amounts to work with.

Recombinant DNA techniques have been enormously helpful in working out the sequences of genes, learning about their functions, and producing biomolecules with valuable medical uses. They have also provided new ways of discovering the genes associated with particular genetic disorders and testing for carriers of these genes. One way is with *gene*

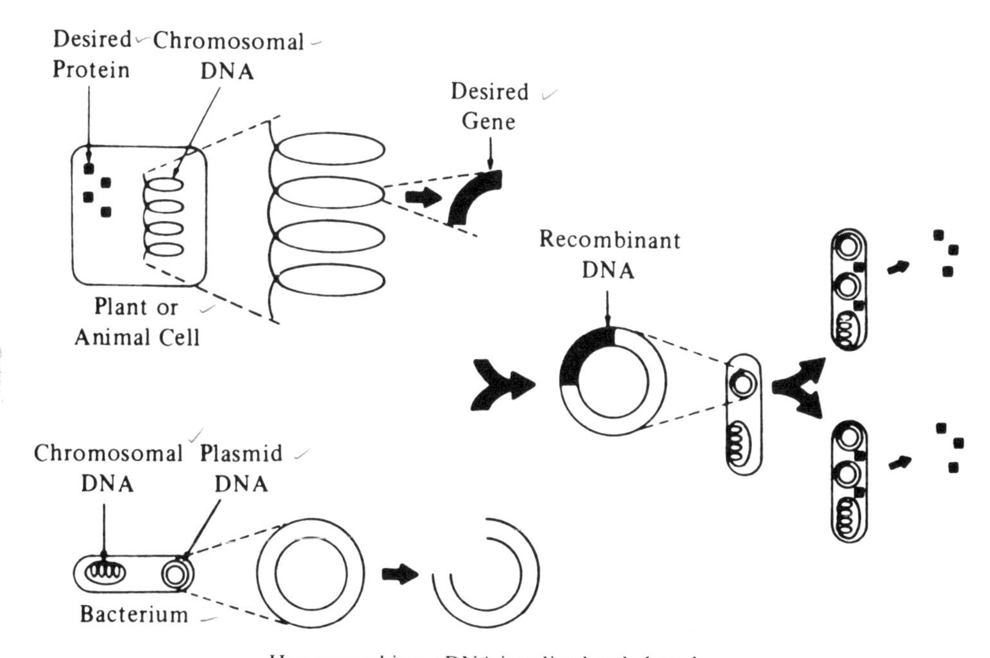

How recombinant DNA is spliced and cloned.

probes. A copy of the normal gene is isolated or synthesized, along with a copy of the mutant gene. DNA from a person's blood cells is treated with appropriate enzymes, then mixed with the gene probes. If the person has only normal genes for the trait, the normal gene probe will exactly correspond to the base sequence of a portion of the person's DNA and will combine, or *hybridize,* with it. But if the person is carrying the mutant gene, it will be the abnormal gene probe that matches the sequence and hybridizes with the person's gene. It isn't even necessary to have isolated the gene for a particular trait in order to make a gene probe; it can be done from a messenger RNA copy of the gene. The mRNA can be isolated from a cell that is actively producing the key protein product, and then an enzyme called *reverse transcriptase* is used to transcribe in the reverse of the usual manner, from mRNA to a DNA copy. This cDNA is single-stranded, and it has the same sequence of nitrogen bases as the original gene. A complementary copy of the cDNA is then made and used as a gene probe.

Recombinant DNA techniques have provided still another way of testing for carriers of mutant genes, even in cases when the gene or its messenger RNA has not yet been identified. When a set of human chromosomes is cut up with restriction enzymes, a characteristic set of fragments of different lengths is obtained. The particular assortment of fragments produced, say, for chromosome 11, will differ from one person to another because there are a number of minor variations in the DNA sequence. Perhaps a few bases have been lost from one spot, or a segment in another place has been duplicated so that it is repeated once, twice, or even more times. Such changes may have no apparent effect on a person's development and health if they happen to fall outside the working parts of the genes. Researchers have named this type of individual variation *restriction fragment length polymorphism,* or RFLP (pronounced "riflip"). They have found that RFLPs are inherited in simple Mendelian ratios, and thus members of the same family will tend to have some RFLPs in common.

Studies of pedigrees of a number of genetic disorders have revealed that certain RFLPs are associated with particular disorders, because they lie on a part of the chromosome close to the gene (or genes) responsible for the disorder. These RFLPs can thus be used as *DNA markers,* indicating a high probability that a person who has them is carrying the mutant gene. This is only a probability, not a certainty, because the RFLP may only be near the gene, not in it, and thus they could have been separated by crossing-over or translocations during cell division.

Another complication in marker testing is the fact that individual families may differ somewhat in the particular RFLPs associated with mutant genes. Genetic screening using RFLPs thus requires having some genetic material from someone in the family who has the disorder, plus samples from normal family members for comparison. This type of test cannot be used, for example, for singer Arlo Guthrie, who may have inherited the gene for Huntington disease from his father, folk-singer Woody Guthrie. Both Arlo's father and his mother died before marker tests for this disorder were developed, and none of their DNA was preserved. Their particular RFLPs, therefore, could not be determined.

More than 100,000 Americans share Arlo Guthrie's dilemma. They have a family history of Huntington disease and do not know whether they have won or lost in the fifty-fifty chance of inheriting the mutant gene. Living with that kind of threat through one's young adult years can be agonizing. Picture watching yourself constantly for the first telltale signs of memory loss or loss of body control. You may be afraid to make long-term career plans, to commit yourself to a serious emotional relationship, and especially to have children. Now, at last, for many children of Huntington victims there is a possibility of answering the question, of knowing what kind of future they face.

Would you choose that possibility?

Before RFLP tests for Huntington disease were available, in surveys of people at risk, between 60 and 80 percent said they would want to know whether they are carrying the gene. But when trial screening programs were set up at Johns Hopkins Hospital in Baltimore and at Massachusetts General Hospital in Boston, the initial response was cautious. In the first six months, only 70 out of 350 people at risk notified by Johns Hopkins signed up for the tests; out of 1500 at risk in the New England area served by the Massachusetts General program, only 32 went in for preliminary counseling during that time. Studies have shown that people at risk have a high rate of depression and behavior disorders; the chronic uncertainty can be draining, yet how well will such people handle the certain knowledge that they will develop the disease if the test shows they have inherited the gene? Columbia University researcher Nancy Wexler, who helped to develop the marker test and whose mother died of Huntington disease, says, "Suicide is not unreasonable." The screening programs are proceeding slowly and cautiously, with emphasis on adequate counseling and emotional support for the participants.

For a disease such as Huntington, carrier screening can provide not only information useful for planning a family but also a key to one's personal fate. But most genetic disorders are not determined by dominant genes (nor so late in onset). For recessive disorders, carrier screening programs for high-risk groups can be extremely effective, saving people from unnecessary grief. But when they have not been thought out carefully, they may actually be harmful. A classic contrast is provided by screening programs set up for Tay-Sachs disease and sickle-cell anemia.

In the early 1970s a mass screening program for Tay-Sachs carriers was set up on a trial basis in the Baltimore, Maryland, area. After intensive publicity to inform Jewish people of the area about the disease and its danger to children, centers were set up to take blood samples for testing. More than 60,000 couples were tested, and a number of carriers of the Tay-Sachs gene were identified. They were called in for genetic counseling, where they learned that if only one parent is a carrier, none of the children will be at risk for the disease (though there is a 50 percent chance that each child will be a carrier). If both parents are carriers, there is a one-in-four chance of having a child doomed to death at an early age. For such couples who decided to take the risk and have a child, amniocentesis could let them know in advance whether their child would have the disease by testing the fetal cells for the enzyme hex A. The pilot program was so successful that it was expanded. By 1981, a total of 350,000 young Jewish adults had been screened voluntarily in 102 centers throughout the world. In North America there has already been a 65 to 85 percent drop in the rate of appearance of new Tay-Sachs cases as a result of carrier screening and prenatal diagnosis.

Screening programs for sickle-cell anemia, also begun in the early 1970s, were far less successful. After screening tests were devised, the media focused on sickle-cell anemia as a "neglected disease," and communities quickly launched sickle-cell screening programs. Unfortunately, many of these programs were poorly planned, with more input from politicians than from medical experts. (A few states even passed laws requiring sickle-cell tests for newborn infants, school children, applicants for marriage licenses, and prison inmates.) Little attention was given to educating the community about the nature of the disease and what the tests meant.

Well-meaning amateurs went around the neighborhoods with testing kits and left people with the knowledge that they were "carriers" but with little idea of what that meant or what to do about it. Often the carrier status (which does not usually involve any increased health risk) was confused with having sickle-cell disease (which can be fatal). Insurance companies began charging sickle-cell carriers higher rates, and six major airlines refused to hire any carriers of the trait because of their supposed health risk. One of the sorest points was the fact that at the time, no method of prenatal diagnosis was available, so that a couple who were both carriers had no way to avoid the possibility of giving birth to a child with the disorder other than to refrain from having any children at all. Black leaders began to see the whole carrier screening program as a racist plot. Many of the poorly planned programs were eventually abandoned. Later, when RFLP tests became available for prenatal diagnosis, better organized voluntary screening programs were set up.

Prenatal Diagnosis

Some of the same kinds of tests that are used for carrier screening, such as karyotyping, tests of enzyme levels, and gene probe and marker tests, can also be used in prenatal diagnosis to determine whether a child will be likely to suffer from a particular genetic disorder after it is born. But how can genetic specialists obtain cells of a developing fetus to test when it is safely enclosed inside its mother's womb? The technique that is used most commonly is *amniocentesis.* Its name comes from the *amnion,* the membrane that encloses the developing fetus and the fluid in which it is suspended.

In amniocentesis, a long hollow needle is inserted through the mother's abdomen into the uterus, and a sample of amniotic fluid is carefully drawn out. This fluid usually contains some stray cells shed by the fetus. These cells are cultured—

grown in dishes with nutrient broth—so that a sample of dividing cells can be obtained for karyotype studies. Amniocentesis is usually performed after the fourteenth week of pregnancy, when the uterus is large enough. There is a small risk, about 1 in 200, that amniocentesis will induce a miscarriage, and there is a very slight risk of infection to the mother, so this procedure is not used for general screening. It is recommended for mothers who have some increased risk: those who are thirty-five or older, or who have had a child with a genetic disorder, or who have some family history of a disorder.

Chorionic villus sampling (CVS), a relatively new prenatal testing technique, can be performed earlier in the pregnancy, after nine to twelve weeks. A catheter is inserted into the uterus through the vagina, and a sample of cells is sucked up from the *chorion,* another of the fetal membranes. The chorion contains rapidly dividing cells, so chromosome analysis may often be completed in just a few days, without having to wait for a cell culture. This technique is still being tested, but the risk of miscarriage in CVS is thought to be just a little higher than with amniocentesis.

Fetoscopy permits the doctor to see the fetus directly by means of a thin fiberoptic probe inserted into the uterus. Samples of fetal blood or tissues can also be taken. This is a somewhat riskier procedure and has about a 5 percent chance of resulting in miscarriage. Only a limited number of testing centers perform it.

Ultrasonography is much safer, with no known risk to the mother or fetus at all. High-pitched sound waves (beyond the range of human hearing) are beamed into the mother's uterus, and the pattern they form, bouncing off structures inside, is interpreted by a computer in the form of a picture on a screen. On this *sonogram,* the outlines of the fetus can be seen (even distinguishing its sex), and some kinds of developmental abnormalities can be spotted. Skilled interpreters of the images can make out telltale signs of Down syndrome: thigh bones that are slightly shorter than usual and an extra roll of skin on the back of the neck. Although the test is not reliable enough by itself for diagnosis of the condition, it may be useful for picking out women who should have a follow-up study by amniocentesis or chorionic villus sampling. Ultrasonography is also used during amniocentesis to locate the position of the fetus and the placenta, so the sampling needle or catheter will not do any damage.

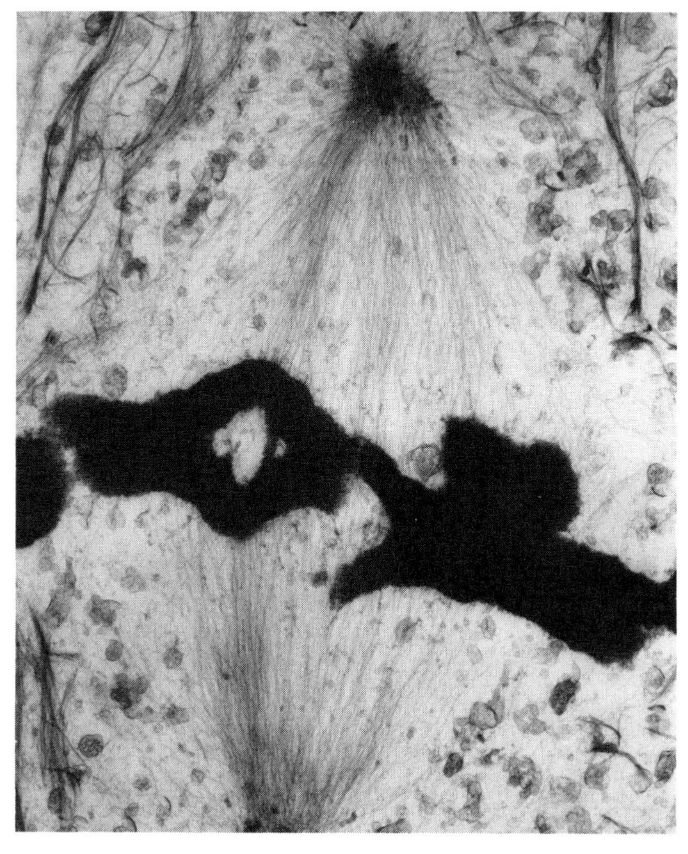

Kangaroo cells, which are quite large, are often used to study rapidly dividing cells, such as those contained in the human chorion. (The big blobs are chromosomes, shown at the stage of mitosis when the spindle is forming.)

Some special tests can also provide information for prenatal diagnosis. A substance called *alpha-fetoprotein,* for example, is normally present in the blood in very low amounts, but during pregnancy a woman's serum alpha-fetoprotein level rises. This protein is produced in the yolk sac of the fetus and passes into the amniotic fluid, reaching a peak there at about twelve to fourteen weeks. In the presence of *neural tube defects,* which are abnormalities of the development of the nervous system, the alpha-fetoprotein levels are unusually high. Tests for this substance in the amniotic fluid can thus point to such birth defects as anencephaly (a lack of brain development) and spina bifida (a failure of the neural tube to close off, leaving an opening at the end of the spinal cord and often resulting in mental retardation). An unusually low alpha-fetoprotein level, on the other hand, can be an indication of Down syndrome. Testing the mother's blood for alpha-fetoprotein levels can also be used to screen for these disorders at about nine weeks of pregnancy. Although the results are not as reliable as tests of the fetus, this simple blood test can be used to screen for neural tube defects; women whose test is positive are referred for amniocentesis and further studies. Recent research in Britain suggests that the addition of other simple blood tests can make screening for Down syndrome even more effective. One of these measures the level of *human chorionic gonadotropin,* a hormone produced during pregnancy; its amount is an average of twice as high in women carrying a child with Down syndrome as in those whose fetus does not have the abnormality. The use of such tests can greatly reduce the need for amniocentesis and CVS and thus cut the risk of aborting a normal pregnancy.

Determination of the sex of the developing fetus can be of value in predicting the risk of sex-linked disorders for which more precise tests are not yet available. A female fetus generally will not have an X-linked disorder, although she may be a carrier. Sometimes, if chromosomes from family members are available for comparison, distinctive markings or structural characteristics may make it possible to distinguish which X chromosome a male fetus has inherited from his mother and whether he will be normal.

Some people object to prenatal testing on moral grounds, believing that people who have such tests will choose to abort any child that is "imperfect." In practice, however, prenatal diagnosis is followed by abortion in only about 2 percent of the cases. Usually the tests reveal that the suspected genetic problem has not occurred. When an abnormality is found, the early detection may permit doctors to treat it effectively—perhaps even before birth. The availability of testing permits many couples to attempt a pregnancy when without it they would have been too afraid. And if such a couple does find that a fetus is suffering from a serious genetic disorder, chooses abortion, and later conceives a normal child—there are many doctors and philosophers, as well as ordinary laypeople, who believe that an ethical choice has been made. It can be a difficult dilemma.

Newborn Screening

At birth a baby with phenylketonuria (PKU) seems normal. During pregnancy, while the baby still depends on its mother's body for nourishment, her phenylalanine hydroxylase passes through the placenta and keeps the baby's phenylalanine level fairly normal. But after birth a baby's metabolism is on its own. A baby with PKU, lacking the proper enzyme, is unable to process the phenylalanine in the protein-rich milk it drinks. Quickly the phenylalanine level in its blood rises. Eventually there is so much of the amino acid in the blood that it spills over into the urine; but there may not be enough for a urine test to be positive until a month or so has passed. Now the screening of newborn infants for PKU is usually done by testing a drop of blood taken from the baby's heel. Even so, the test results may not be accurate. Generally the cut-off point for considering a PKU test positive is set at a very low level, to minimize the chances of missing any babies at risk. This gives a number of false positives, but they can be eliminated (after quite a bit of worry by the parents) by follow-up tests at the baby's first visit to the pediatrician or clinic.

In the excitement that followed the development of reliable tests for PKU screening in the early 1960s, one state after another rushed to pass laws requiring testing of all newborn infants. (Virtually all of the babies born in the United States are now routinely tested for the disorder.) The laws make sense, even though PKU is a rare disease, because when it is detected early enough, mental retardation can be prevented by feeding the child a special diet free of phenylalanine. Children on the diet grow normally and remain healthy.

As the years went by, though, some problems began to emerge. Patients found the PKU diet very "blah"-tasting and boring. When they were very young, they ate what their parents told them to. But then when they went to school and saw all their friends eating forbidden treats like ice cream, hamburgers, and pizza, they began to rebel. Gradually doctors decided to take children with PKU off the special diet at the age of about six, hoping that by that time their intelligence had developed sufficiently. At first this seemed to be true, but then it was found that after years on a normal diet the high phenylalanine levels cause some loss of intelligence in people with PKU, although their mental abilities do not fall to retarded levels. Then another problem surfaced. As the first girls with PKU raised on the special diet grew up and began to have children of their own, it was discovered that high phenylalanine levels in a mother's body can damage the brain of her developing child. Now doctors advise a woman who has been treated for PKU to go back on the special diet before becoming pregnant and to keep her phenylalanine intake low during pregnancy.

Despite the problems, PKU screening seems to be a resounding success. It has been estimated that the entire PKU screening program in New York State, for example, costs about $250,000 a year—considerably less than the cost of lifetime care in an institution for a few mentally retarded victims of the disorder. In dollars and cents, that seems like a real bargain—and who can put a value on the saving of so many minds that would otherwise have been wasted?

In addition to PKU, New York State now requires newborn screening for sickle-cell anemia, congenital hypothyroidism (a lack of sufficient secretion of thyroid hormone, which helps to regulate the rate at which the body uses food materials), and some very rare metabolic disorders, including maple syrup urine disease, homocystinuria, histidinemia, galactosemia, and adenosine deaminase deficiency. In disorders such as hypothyroidism, as with PKU, early treatment by modifying the diet can prevent the symptoms of the disease from developing. (A child with insufficient thyroid hormone becomes mentally retarded and physically stunted; yet doses of thyroid extract given early enough can permit a child with hypothyroidism to develop normally.)

In the case of sickle-cell anemia, newborn screening makes good sense, too. Studies have shown that giving sickle-cell children oral penicillin prevents devastating bacterial infections. Up to the age of about three, a child with sickle-cell anemia has difficulty fighting off such infections; a strep infection can develop so quickly that a child may seem perfectly well only six hours before death. In California, children with sickle-cell anemia are given penicillin and inoculated with vaccines against several dangerous infections. Their parents are given careful counseling, with follow-up family education sessions that teach them to recognize emergency signs.

The experiences of a New York mother, reported in a study by the National Institutes of Health, point up the need for such counseling. The mother was not informed of the results of the sickle-cell tests routinely done on her newborn baby. She did not find out that her daughter suffered from the disorder until after she had twice taken the baby to the emergency room of her local hospital. Each time the doctors on duty failed to recognize that a fever, swollen joints, and an enlarged spleen were signs of a sickle-cell crisis and instead accused the mother of child abuse.

The benefits of discovering that a child has histidinemia are unclear—this metabolic "disorder" has no symptoms that anyone has yet discovered. For maple syrup urine disease, screening raises difficult ethical dilemmas; at present, costly treatments may only delay death by a few years, which are filled with suffering for the child and its parents.

The study of boys with XYY syndrome mentioned in the last chapter raises another ethical question: Do such studies, designed with the expectation of detecting abnormalities because of the abnormal karyotype, turn into self-fulfilling prophesies? Were the boys found to be impulsive and aggressive because that was what the researchers had expected to find? Or did their parents treat them differently than they would have because they felt there was something wrong with them genetically? Again, careful counseling of the parents is vital in such cases so that the awareness of individual differences can be used to help children to fulfill their best potential.

Genetic Counseling

Typically, people seek genetic counseling after they—or a family member—have had a child with a genetic defect. They want to know more about the defect and what the chances are that later children might inherit it too. Women thirty-five or older who are pregnant or planning pregnancy may also be advised to seek genetic counseling or screening. People who have a genetic disorder or have been found to be carrying a gene for a disorder in a general screening program are also in need of genetic advice. Others who come for counseling include couples who are related to each other (cousins, for example), members of particular ethnic or racial groups, and women who have had repeated miscarriages or who have been exposed to a teratogen during pregnancy. (In some areas, special "teratogen hotlines" have been set up to give advice about the risk of birth defects after exposure to drugs, chemicals, radiation, or to illnesses such as rubella.) Sometimes a couple planning to adopt a child or an agency trying to place a child for adoption will need advice about the implications of a genetic disorder.

Genetic counseling is provided by hundreds of counseling units, located mainly in regional medical centers and teaching hospitals. The counselors are expert professionals and include not only doctors and geneticists but also specially trained nonphysicians who have completed a master's degree program of

Looking at a karyotype during a genetic counseling session.

study in genetic counseling.

The genetic counseling process usually begins with taking a detailed *family medical history*. Information about disorders in family members may be checked by running blood tests or chromosome analyses. Sometimes that is not possible if family members refuse to cooperate or key members have died. Doctors today are advised to perform an autopsy on any child who was stillborn or died during the *neonatal* period—shortly after birth. In addition to a description of any abnormalities found during the autopsy, chromosome analysis, photographs, and X-rays may be helpful to the genetic counselor in establishing a correct diagnosis. Information on the nature of the disorder provides the basis for sound counseling, assessing the risk for future children of the couple.

Estimating the risk is the next step for the genetic counselor, who arrives at a decision on the basis of a study of the pedigrees, test results, and any pertinent medical publications. Often the conclusion is a happy one for the parents. It turns out that the disorder from which a previous child or family member suffered was not genetic after all, or that it was a random event—something that could happen to anyone, with no special likelihood of recurring. If that is not the case, the genetic counselor may be able to calculate the odds fairly precisely. Even if the condition does not follow strict Mendelian laws, there may be enough past experience with the same kind of disorder to calculate probable risks.

Communication is the next task for the genetic counselor: explaining to the couple exactly what kind of genetic risks they face, what the odds are, and what options they can choose to avoid giving birth to a seriously genetically handicapped child. If the genetic condition can be treated, the parents will need full information about what the treatment entails, both in cost and in emotional stress, and what its prospects are for success. If no acceptable treatment is available, the couple needs to know what preventive measures they can take. This information may range from material on contraceptive techniques to a description of alternative birth options.

The genetic counselor does not make a decision for the couple; only they have the right to do that. The role of the counselor is to make the options clear so that the couple can make an informed decision. The counselor may also provide information about the various support services that are available for couples facing such difficult choices, with referrals to appropriate social workers, clergy, or medical specialists.

Today's biomedical science and technology are providing alternatives that were unavailable to any past generation. National Institutes of Health bioethics specialist John Fletcher has complained that there are now twenty-four ways to have a baby, and the choices are "morally dizzying." The new *birth technology* options have been devised mainly to help couples with problems of infertility, but some of them are of special interest to couples with genetic risks.

Artificial insemination—donor (AID) is the oldest of the birth alternatives. Semen from a male donor is inserted into a woman's vagina. Normally this is performed by a physician, using semen from a donor who is kept anonymous, but some women make private arrangements and may even insert the semen themselves, using a sterilized turkey baster. The insemination is carefully timed to a woman's most fertile time of the month, but it typically takes an average of several tries before a pregnancy is achieved. AID can be a valuable alternative if both members of the couple are carriers of a harmful gene. It has been estimated that by 1980, about one third of artificial inseminations in the United States were already being done for genetic reasons. AID is not a foolproof solution though, particularly since semen donors are not usually screened for genetic disorders.

A *surrogate mother* bears a child for a couple who cannot have one in the normal way or should not for medical reasons. The surrogate may be artificially inseminated, or the baby may be conceived *in vitro*—a "test-tube baby" (who should actually be called a "petri dish baby," since the culture in which the embryo develops at first is in a flat-bottomed dish rather than a test tube). Ova in a woman's uterus are stimulated to mature by hormone treatments and then collected for the in vitro fertilization. Another variation that could help a woman carrying a damaging gene is *ovum transfer,* in which an ovum from a healthy donor woman is fertilized with the husband's sperm and then placed into the uterus of the wife or a surrogate mother.

Freezing provides some further options. Sperm can be frozen and stored for later use in artificial insemination; ova and even fertilized eggs or very early embryos can also be frozen and later implanted. If you'd like to choose the sex of your child, there are seventy-five clinics in the United States that promise to help you conceive a boy or a girl—at a fee of $600 for a male and $500 for a female. Hypothetically, cells from a very early embryo could be split off and developed separately, producing twins, triplets, or larger numbers of genetically identical babies. In experiments in animals, multiple genetic copies of a single organism—*clones*—have been produced by inserting the nuclei of cells from an embryo into ova whose own nuclei have been inactivated. Although claims of similar achievements in humans are believed to be only science fiction at this point, it is likely that cloning of humans will eventually be feasible. It may even be possible some day to turn the developmental genes of cells from an adult back on and produce cloned copies of someone who already exists. (The adult and his or her cloned offspring would share the same heredity, but they would not be identical: In addition to the difference in their ages, differing environmental conditions while the clones were growing up might introduce numerous variations.)

We will leave you to imagine the ethical complications of meeting a few clones of yourself walking down the street; the real ethical dilemmas that exist today are difficult enough. For example, processes of in vitro fertilization typically start with a dozen or more ova. Not all of these are fertilized, but usually there are three or more embryos growing in the culture dish when the time comes for insertion into the woman's uterus. Doctors prefer to insert several embryos because doing so increases the chance that at least one will implant and develop successfully. Usually all but one die off—which means a loss of life if you believe that unique human life exists from the moment of conception. If more than one embryo implants, as sometimes happens, a multiple birth may result. But conditions are crowded in a uterus that holds more than one developing fetus, increasing the chances of abnormalities. And what about the extra embryos that may have been left over in the dish? Should they be thrown away? Or frozen for possible later use? If so, to whom do they belong? To their biological parents? To the client parents if donor ova or sperm were used? What if the parents die while the embryos are still frozen?

Some more dilemmas: If the testing laboratory makes a mistake and a child who was supposed to have been healthy turns out to have a genetic disorder, who is at fault? Can the parents sue the genetic counselor for malpractice? Can the child sue for "wrongful life"? Such suits have already been filed. What about the question of abortion? The *Roe* v. *Wade* decision by the United States Supreme Court established that women have a legal right to abortion, but opinions on the ethics of the practice are sharply divided. Polls generally show that the majority feel an abortion—at least in the early stages of pregnancy—is a legitimate choice for a woman to make. But many people sincerely believe that abortion at any stage, for any reason, is murder; and they are trying to change the laws to make it illegal. If you do not go along with that view, then at what stage does a fetus have rights? If prenatal diagnosis and treatment for a genetic condition are available, do the parents have an ethical right to have an abortion? Does the fetus have a right to treatment? What if the treatment is risky for the mother?

Doctors, clergy, and lay people are now wrestling with many thorny dilemmas like these. Some people believe we need laws to regulate the various practices of genetics and birth technology—perhaps even to ban some of them. Others think that these are private matters of individual conscience, and the government should not intrude. Meanwhile, disputes are being brought to the courts, and precedents are being established. The controversial Baby M case in New Jersey, for example, wrestled with the thorny conflict pitting a childless couple who contracted to have a child conceived by artificial insemination and borne by another woman against the surrogate mother who changed her mind and wanted to keep the child. The final decision, while awarding custody of the child to her biological father and his wife, preserved the visiting rights of the biological mother and raised questions about the legality and advisability of such birth technology practices. This and other controversies are still far from resolved.

Genetics in the Workplace

The applications of biotechnology to genetic screening are bringing new choices—and new dilemmas—to couples who would like some control over their reproductive future. These same tests, gene probes and DNA markers, can also be applied to adults, to determine their risks for various diseases.

Such tests can play a valuable role if used to help people minimize these risks—for example, by avoiding overexposure to the sun if they have a hereditary sensitivity to ultraviolet rays, or by switching jobs if they are particularly sensitive to chemicals used in the workplace. Even so, some people would prefer not to know what genetic defects they may harbor.

To test or not to test may not always be a matter of individual choice. Insurance companies want to use the new genetic tests to fine-tune their actuarial tables that predict the life expectancies of various population groups. Employers, thinking about promotion tracks and retirement programs and conscious of "cost effectiveness," may try to force employees or job applicants to take genetic tests. Is this legal? Is it moral? Does anyone but the individual have the right to know the details of his or her medical condition—or possible future condition? Do the rights of privacy outweigh responsibilities to the community? No one is quite sure of the fine points of all these questions yet, and the various states are just beginning to grope their way toward legal solutions.

There are humane and beneficial ways to use genetic screening results. But will we make these wise choices?

Allowing insurance companies to require genetic tests, just as they are currently allowed to require a medical examination and the disclosure of past medical history, will mean that some people will have to pay higher insurance rates because they have genetic disorders that increase their health risks. Many people's rates will go down because they have fewer genetic risks than the average or because they are taking special precautions to stay healthy, guided by their genetic profiles. Insurance companies already offer reduced rates as an incentive to nonsmokers. But there are also precedents for abuse. Some insurance companies have been refusing to write policies for people in high risk groups for AIDS, or demanding that applicants take AIDS antibody tests, even though no one is yet sure how many of those whose tests are positive will eventually develop the disease. Some fear that genetic tests may be used in a similar way and that people with genetic health risks may be unable to buy insurance at any price. Yet if they get sick, someone has to pay for caring for them, and that may mean higher taxes for all of us.

Another worry is that insurance companies may not keep genetic test results confidential, and these results may be used as a basis for discrimination against the "genetically unfit," just as the sickle-cell screening results were unfairly used to deny jobs or insurance to carriers of the trait.

Some employers are already trying to use genetic screening in a creative and positive way. In a "wellness" program for one thousand employees at Chesapeake & Potomac Telephone, computers analyze family medical histories, and blood samples are analyzed for fourteen diseases, including cervical cancer, diabetes, and gum disease. A "health advocate" goes over all the test results with the participant, who signs a "personal health action plan" outlining preventive steps to reduce health risks. Participation in the plan is voluntary, but as genetic tests become more widespread, some companies may make testing compulsory, just as some employers now require AIDS tests or drug tests or subject employees to routine lie detector screening. Four states now have laws banning genetic testing of employees. Florida, Louisiana, and North Carolina prohibit sickle-cell testing. In New Jersey, all forms of genetic testing by employers are illegal, as is discrimination on the basis of genetic test results. As the use of genetic testing grows, more decisions will have to be made on how they may be used.

6

Treatment: Today's Options

If you wear glasses or have had your teeth straightened by braces, you are living proof that genetic disorders can often be treated successfully without changing the genes responsible for them. Your heredity has created a disability—perhaps a set of teeth that don't meet in efficient alignment for biting and chewing, or eyes that cannot focus well on objects at a convenient distance for reading. Such problems can be solved by physically changing the faulty body parts (moving the teeth into a better alignment with orthodontic appliances) or changing the environment to compensate for the hereditary lack (placing corrective lenses in front of the eyes to help focus the incoming light rays).

A number of genetic disorders can be treated with approaches like these. Birth defects such as clubfoot, cleft lip and cleft palate, and congenital heart defects, can be corrected surgically—literally repairing the defective body part. As researchers learn more about the biochemistry of the body and its complex systems, doctors are also gaining the ability to compensate for missing or faulty gene products with changes in diet, with drugs, or with replacements for the lacking biochemicals. Treatments for diabetes, thyroid deficiency, and hemophilia have been available for a long time. Now recombinant DNA technology and new drug delivery techniques are permitting doctors to tackle other genetic disorders as well.

Prenatal Surgery

When a baby is born with an abnormality such as a misplaced heart valve or eyes that do not focus together, the doctor may advise putting off corrective treatments for a while, until the baby is larger and better able to tolerate the surgery. But some disorders can't wait—they hamper the child's development in other ways, creating more serious problems. Sometimes it is already too late when the baby is born. A herniated diaphragm, for example, is an easy problem for a surgeon to correct. It is just a matter of closing up the hole in the diaphragm, the dome-shaped muscle that forms the floor of the

Identifying gene products using autoradiography.

chest cavity, so that organs from the abdomen can't push up through it. But a newborn baby is already nine months old. It has been growing and developing inside its mother's uterus all that time. If its abdominal organs were pushing up through a hole in the diaphragm, they were decreasing the space inside the chest cavity, so that there was not enough room for the lungs to develop properly. Without well-developed lungs, a baby cannot take in enough oxygen from the air after it is born and must breathe on its own. Such babies often die shortly after birth.

Microsurgery has been making steady advances. Surgeons can now routinely perform operations on newborn infants that used to be put off for months or even years. Recently they have begun to act even more boldly, opening the uterus to operate on a tiny one-pound fetus and then returning it to the womb to allow it to continue its development. Prenatal surgery has been especially successful in cases of *urinary blockage,* which causes urine to back up in the fetus's abdomen. Severe kidney damage, malformations of the lungs and legs, and even death may result. But after an operation to drain the fluid and correct the blockage, the fetus can develop normally. Physicians hoped that prenatal surgery could produce just as dramatic a cure in the case of *hydrocephalus,* a buildup of fluid in the brain that may swell the head so much that birth is difficult, meanwhile crowding the delicate brain tissue and stunting its development. So far, however, the results have been disappointing. Although the surgeons are able to eliminate the "water on the brain" by inserting tiny drainage tubes, about two-thirds of the babies that survive the operation are born mentally retarded.

The field of prenatal surgery is still new. The first such operations were not performed until 1981. It is an area in which the pioneers must make difficult decisions. When the uterus is opened to operate on a fetus, there is a substantial risk of causing a premature birth. Therefore, the doctors hope to wait until the fetus is old enough to survive on its own if such a mishap occurs. But if they wait too long, the damage may no longer be repairable, or the fetus may even die. The specialists must evaluate a fetus very carefully before deciding on surgery, since the problem that shows up on a sonogram may not be the only one. The surgery may be successful, but the baby may turn out to have other defects so serious that it cannot survive.

Who has the right to make the final decision on whether a fetus with a severe but correctable defect should be treated in the womb? At present this is the right of the parents; the role of the doctors is to provide information about the options and their risks and consequences. There have already been legal challenges to the parents' rights to approve or withhold treatment of newborn infants with serious abnormalities, and the legal and ethical status of the fetus is unclear. Does a fetus have rights? Should it be treated even if this would endanger the health of the mother? These are questions that seem certain to arise more and more often as medical knowledge and skill advance.

Biochemical Manipulations

Some biochemical treatments of genetic disorders have already been described: thyroid extract to prevent mental and physical retardation in the case of thyroid deficiency, for example, and the special phenylalanine-free diet that permits PKU children to develop normally. People with hemophilia are treated with clotting factors obtained from processed donor blood. Several different forms of the "bleeders' disease" can be treated with various factors that contribute to the complex chemical sequence of clotting reactions. With these replacements for their genetic deficiencies, these people need no longer fear that an accidental cut or a tooth extraction might result in a life-threatening hemorrhage.

Like thyroid deficiency, *diabetes* is a disorder that involves a hormone. *Insulin,* produced in the pancreas, causes sugar in the blood to be taken up by the body cells and the excess stored in the liver in the form of the animal starch, glycogen. Insulin-dependent diabetes, which typically first appears in children, teenagers, or young adults, is the result of an inability of the pancreas to secrete enough insulin in response to the rise in blood sugar following a meal. If the blood-sugar level soars high enough, some of the excess spills over into the urine; at very high levels coma or even death can occur. Maturity-onset diabetes is a more complex condition. Insulin is produced, but it does not work properly, perhaps because it is broken down too quickly by enzymes, or its action is counteracted by another pancreatic hormone, glucagon, or there are not enough working insulin receptors on the surface of the cells. The insulin-dependent form of the disease can be treated by replacing the lacking hormone. Usually insulin from cattle or pigs, very similar to the human hormone, is used to treat diabetes. Unlike thyroid extract, insulin must be injected; if it were taken by mouth, the digestive juices would break it down before it could do any good.

Insulin injections are carefully calculated and timed to keep the blood-sugar level as steady as possible. With them, a diabetic can live a fairly normal life. Still, many diabetics develop complications, which may include nerve and blood vessel problems and blindness. One cause may be a body reaction against the cattle or hog insulin, which is not exactly the same as the human hormone. Recombinant DNA synthesis of human insulin is providing an alternative that may solve this problem. Another possible reason is the fact that injecting a large amount of insulin into the body all at once is not the way the normal body works. The pancreas secretes its hormone in tiny amounts, whenever signals of a high blood-sugar level tell it that more insulin is needed. Automatic insulin pumps, which deliver tiny measured doses, are now available to help smooth out the insulin delivery. Tabletop, portable, and even implantable insulin delivery systems are on the market. Researchers are working on a true implantable artificial pancreas, which will include not only a delivery pump but also a sugar sensor and a tiny computer chip that will automatically calculate the amount of insulin needed. Transplants of insulin-producing pancreatic cells may also help diabetics in the future. Researchers are currently working out ways to get the cells established in the body and to protect them from rejection by the body's immune defenses.

Recombinant DNA synthesis is also providing human growth hormone for the treatment of dwarfism resulting from a deficiency of the pituitary growth hormone. (This is quite different from achondroplastic dwarfism, a connective-tissue disorder that leads to shortened limbs with a normal-sized head and torso. *Pituitary dwarfism* is due to a slowing of the growth rate and produces a normally proportioned midget.) Early treatment of pituitary-deficient children with growth hormone permits them to grow to a normal height and to mature sexually. Before the recombinant DNA hormone was available, the treatment was extremely expensive. Natural growth hormone could only be obtained from the pituitary glands of people who had died, and a single year's treatment required 150 pituitary glands!

In metabolic diseases associated with an enzyme, there are several ways to attack the problem. Enzymes catalyze

Analyzing the effectiveness of new pharmaceutical products through recombinant DNA technology.

chemical reactions. When a key body reaction is stopped, a needed product may not be formed, or a substance that should be broken down may instead build up and poison the body. The PKU diet approaches the problem by eliminating the source of the substance normally broken down by the lacking enzyme. Another metabolic disorder that can be treated in this way is *galactosemia,* in which a deficiency of one of the sugar-regulating enzymes, galactose-1-phosphate uridyltransferase, results in mental retardation, eye cataracts leading to blindness, and liver disease. Feeding a galactosemic infant a special milk formula without galactose completely prevents all these problems.

In Lesch-Nyhan syndrome, the metabolic defect is a deficiency of hypoxanthine-guanine phosphoribosyltransferase (HPRT), a key enzyme in the reactions leading to RNA and DNA synthesis. One of the results of its absencc is a buildup of uric acid and severe kidney damage. Children with Lesch-Nyhan syndrome can avoid serious trouble by taking allopurinol, a drug that prevents the accumulation of uric acid. In another metabolic disorder, *Wilson's disease,* high levels of copper in the blood result in progressive damage to the brain and liver. This damage can be reduced by treatment with penicillamine, which lowers the copper level.

Enzymes typically work with vitamins and other cofactors. In some metabolic diseases the enzyme is present, but a cofactor is lacking, or the interaction with it is not efficient enough. The B vitamin *biotin,* for example, is necessary for a number of metabolic reactions, catalyzed by several related enzymes. In some infants the body's use of biotin is disrupted in some way. The result is a loss of hair, listlessness, coma, and susceptibility to infection. In one case, a baby girl was born with the condition and died only a few days after birth. Later her parents had another child with the same metabolic disorder.

Fortunately, he was taken to the same hospital that had tried to save his sister. This time, with the experience the staff had gained, they knew what to do immediately. Since there was apparently a problem in the body's use of biotin, they gave the baby a supplement of large amounts of the vitamin, hoping that the extra supply would meet his body's needs. He recovered promptly and stayed healthy on a biotin-supplemented diet. Now biotin is the standard treatment for this metabolic disorder not only after birth but even before it. In a case reported by doctors at the University of California in San Francisco, amniocentesis at sixteen weeks (performed because the mother had previously borne a biotin-dependent child) showed that the new baby would have the disease, too. Reasoning that the earlier they started the treatment, the better the baby's chances would be for a normal life, the doctors began to give the mother large amounts of the vitamin, starting in the twenty-fourth week. (A normal pregnancy lasts thirty-nine weeks.) The vitamin passed through the placenta, and the baby was born healthy.

Metabolic disorders in which an enzyme is lacking can be treated by replacing the enzyme. There are a number of problems in this approach, but they are gradually being overcome. First of all, you can't just swallow a dose of an enzyme (unless it is one that normally works in the stomach) because enzymes are proteins, and as soon as they hit the acid and other digestive juices in the stomach, they begin to break down. By the time they get into the blood, which could carry them to the cells where they are needed, they are not enzymes any more—just a mixture of amino acids. Therefore, like insulin, replacement enzymes have to be injected. Another problem is that if the body is not making any of the enzyme, the immune system will regard it as a foreign chemical and may produce antibodies to fight it. An allergy-like reaction results, and the effects may be serious.

Researchers are working out ways to get around this difficulty. One way is to enclose the enzymes in tiny *microcapsules* (or *liposomes*) made of a fatty substance similar to the membranes of cells. Usually the chemicals *(substrates)* on which enzymes work are rather small molecules. They can pass through minute pores in the liposomes and be acted on by the enzymes while those foreign proteins are safely locked in, protected from the body's defenses. A similar effect is produced by binding the enzymes to studs of a waxy substance called *polyethylene glycol* (PEG). The PEG studs block out the immune cells while allowing the substrate molecules to slip through.

Limited success has been obtained with enzyme treatments in *Gaucher's disease,* in which the lack of an enzyme of fat metabolism results in damage to the liver, spleen, and brain, and *Fabry's disease,* another fat metabolism disorder in which fats slowly build up in the kidneys and prevent them from getting rid of poisonous wastes. Better success has been achieved in a type of *severe combined immune deficiency* (SCID) due to a lack of the enzyme adenosine deaminase (ADA). Children with this disorder do not have the normal defenses against infectious diseases, and an ordinarily minor cold or strep infection can turn into a life-threatening siege. Weekly injections of ADA from cattle, chemically bound to PEG, completely restore the immune systems of children with this disorder.

Tay-Sachs disease is another enzyme-deficiency disorder. The tragic mental and physical deterioration that kills these children within their first few years is due to a buildup of fatty substances in the brain and nerves. Researchers hoped that supplying hex A, the missing enzyme, would help Tay-Sachs children, but there was no improvement. Studies showed the reason: the enzyme was not being taken up by the brain cells, where it is needed. Most chemicals do not pass directly from the blood into the brain tissue. They are screened out by another of the body's defense mechanisms, called the *blood-brain barrier.* There are substances that can pass through this barrier—generally they are those that dissolve readily in fats and thus can get through the fatty membranes of the cells that make up the capillaries, tiny blood vessels that form a network through the tissues and bring the cells needed supplies of oxygen and food materials. By attaching drugs or enzymes to fat-soluble carrier molecules, it may be possible to ferry them through the blood-brain barrier. If so, then the damage to the nerve cells of Tay-Sachs children could be prevented.

Having to take injections every day or week can be bothersome. How much better it would be if, instead of supplying doses of missing enzymes or hormones, doctors could give a child with a metabolic disorder a shot of cells that can produce the needed biochemical and continue to produce it. Researchers are already doing just that. Earlier in the chapter we mentioned the experiments on transplanting islet cells from the pancreas, which secrete insulin, to treat the insulin-dependent form of diabetes. A number of genetic diseases involving the blood system can be treated successfully by inserting cells that produce the appropriate enzyme or other product into the patient's bone marrow. The bone marrow must first be specially treated to make way for the new cells, and sometimes there is a potentially serious reaction called *graft versus host disease,* in which the transplanted marrow cells attack the person's body tissues; but the number of genetic disorders successfully treated by *marrow transplants* is growing. These include various forms of severe combined immune deficiency (in which the transplanted marrow supplies the disease-fighting white blood cells that are lacking), Wiskott-Aldrich syndrome (an X-linked condition with a lack of blood platelets and various immune deficiencies), thalassemia (a hemoglobin disorder common in people from the Mediterranean region that, like sickle-cell disease, produces severe anemia), sickle-cell anemia, Fanconi anemia, and Gaucher's disease.

In the sequence from preventing or removing poisonous buildup products, to supplying missing cofactors, to supplying missing enzymes, and finally to supplying cells that produce the lacking enzymes, we have been getting progressively closer to fixing genetic disorders by making the abnormal body systems normal. Why don't we try to go all the way? say some genetic engineers. Let's fix the genes themselves so that they produce the right products, or supply the cells with new genes to do the job. Before discussing the current progress toward true gene therapy, let's explore some of the new tools and techniques of biotechnology that are making gene therapy possible.

7

Tinkering With Genes

In the spring of 1986, molecular biologists meeting at Cold Spring Harbor, New York, boldly proposed a new research project: to determine the complete sequence of the human genome—all of the 100,000 or so genes, with their billions of nucleotides. By the summer, the weather had heated up but the scientists' enthusiasm had cooled a bit. Fearing that the three-billion-dollar investment that such a mammoth project would require would drain funds away from more interesting and important studies, they pointed out that starting such a project at that time would be wasteful and premature, since future technological advances would provide means of obtaining the sequences faster and more cheaply. There was some wisdom in that argument: The "project genome" proposal was based on a dramatic advance that had just been announced by the California Institute of Technology—the development of a new automated DNA sequencer that could read off the order of the nucleotides at a rate of about 8000 bases a day—ten times as fast as the best manual methods available. But by early 1987, Japanese researchers announced that they had developed a sequencer capable of determining 300,000 bases a day (one every quarter of a second!) and planned to set up a "factory" that could run through a million bases a day at a cost of about 10 cents a base. Some molecular biologists have also questioned how valuable it really would be to know the *whole* genome sequence, since about half of the genome is believed to contain repetitions and nonsense sequences—a kind of "molecular junk."

In a later conference the scientists decided on a more modest goal: first to work out a rough physical map of the genome by cutting it into small, overlapping fragments with restriction enzymes and determining their order. Then the precise sequences could be determined for fragments that are particularly interesting because they are involved in genetic disorders or in important body functions. Afterward, the rest of the sequence could be determined at leisure.

Researchers have come a long way since the early days of molecular biology. The complete sequence of the first protein—insulin—was announced in 1954 by British biochemist Frederick Sanger. The job had taken him ten years and won him a Nobel prize. Yet insulin is a rather small protein, containing just 51 amino acids. Today's researchers can use automated protein sequencers to read off, in just a few days, the sequence of a protein containing thousands of amino acids. The first synthesis of an artificial gene, by Nobel prize-winning molecular biologist Har Gobind Khorana in 1970, was another tremendous feat. Now, for less than $30,000, a laboratory can buy an automated "gene machine" that builds up a gene, one nucleotide at a time. All the researcher has to do is type out the genetic code for a particular gene on the machine's keyboard. It takes about half an hour to add each nucleotide.

Early in 1987, researchers at Columbia University announced that they had made the first complete genetic map of a living organism: the bacterium *Escherichia coli,* which lives in human intestines. The map was crude, showing large segments of the *E. coli* chromosome, containing clusters of related genes, but it took eighteen months to compile. Yet *E. coli* has only a single chromosome, which is one-tenth the size of the smallest of the human chromosomes (chromosome 21, with about fifty million base pairs). As for progress on the human genome so far, by the time the *E. coli* map was announced, geneticists had determined the complete sequences for about a hundred human genes (some ten million base pairs out of the three billion in the human genome) and had mapped out the genetic influences of about a thousand genes. With the latest computer methods, scientists can compare a new gene sequence with all those that are already known and

Using a computer to chart the chemical reactions used to decode DNA.

pick out similarities that may provide new insights into the effects of the genes and how they work.

Progress on the creation of a RFLP map has been accelerating. Although the genetic markers yielded by restriction enzymes give only the approximate location of faulty genes, they have already provided the basis for diagnostic tests and are permitting researchers to zero in on key chromosome locations in their search for defective genes. By mid-1987, comprehensive RFLP maps had already been made for human chromosomes X, 7, and 16, and substantial parts of the maps had been filled in for chromosomes 12, 13, 21, and various others.

Meanwhile, in 1988 the United States embarked on "project genome," a government-funded effort to determine the identity, biological function, and ultimately the composition of all the human genes. This program is expected to take fifteen years and cost $3 billion.

Isolating and Cloning DNA

To sequence and identify genes—not to mention using them for gene therapy—researchers must have convenient short lengths of DNA to work with, in a form that permits the different pieces to be sorted out, and in large enough amounts. Isolating pure DNA from cells or tissues is the first step. In work with human chromosomes, DNA is usually extracted carefully from white blood cells, in the form of long double strands hundreds of thousands of nucleotides long.

For several important reasons restriction enzymes are used to chop up these long strands into more convenient-sized pieces, rather than just running them through a blender. When DNA strands are broken mechanically, they snap off straight across, leaving blunt ends. Restriction enzymes, on the other hand, cut the two DNA strands at slightly different points, so that when the segments come apart at the break, each one is left with a short, dangling "sticky end." The sticky ends later prove very convenient for attaching the DNA segments to other pieces of DNA, prepared with the same kind of enzymes. Moreover, each restriction enzyme cuts DNA at a specific point, marked by a particular base sequence that acts as a recognition site, and the DNA fragments formed end with a specific base. Therefore, by using several restriction enzymes with carefully selected restriction sites, a collection of fragments ending in A, C, G, and T can be produced.

The collection of DNA fragments can be separated according to length either by high-speed centrifugation (the different fragments are distributed along the length of the centrifuge tube according to their relative weights) or electrophoresis in a layer of gel (under an electric field, DNA, which has a natural negative charge, is drawn toward the positive end of the gel, with shorter pieces moving faster and farther than longer ones).

The complete DNA isolated from the cells is treated with a number of restriction enzymes to generate a collection of fragments of different lengths. The reaction is stopped before the DNA has been completely digested so that there will not be many very small fragments in the mixture, that at least parts of any particular gene will be included in a variety of different fragments, and that there will be a good chance at least one fragment contains the whole gene (in other words, that there is a fragment in which none of the cuts have occurred inside the gene sequence). The set of DNA fragments for a whole genome thus represents a *genomic library,* the various volumes of which contain the complete set of genes for the organism. Researchers wishing to study particular parts of chromosomes can work with samples of the appropriate fragments (like borrowing books from a library), and samples of sequences that have not yet been identified can be assigned to different workers to avoid duplication of efforts. But first the library must be stored in a convenient form. The method used is cloning, which also produces large numbers of identical copies of each DNA fragment.

A clone is a line of cells all derived from one original parent cell and all possessing exactly the same heredity. Cloning DNA refers to a process of inserting it into a simple, rapidly dividing organism, such as a bacterium, in which it will be faithfully reproduced each time the host organism divides. To achieve this result, the DNA must first be incorporated into a suitable *vector,* which will take care of the tasks of getting into the host cell and directing the cell to copy the new DNA along with its own chromosome. One convenient type of vector is the plasmid, a small circular bit of DNA. Plasmids occur naturally in certain microorganisms, including the well-studied *E. coli,* and they are sometimes transferred from one bacterium to another. The plasmids found in nature generally carry genes that provide some advantage to the host bacterium, such as an ability to survive in the presence of an antibiotic. The circular plasmid DNA can be cut open by treating it with a restriction enzyme and then with a fragment of DNA from another organism—even from a human being—can be spliced in, using DNA ligase. The result is a recombinant DNA molecule, which still looks like a plasmid and acts like a plasmid but is somewhat larger and contains an added foreign gene.

Researchers have isolated a number of natural plasmids and have tinkered with them, putting in or taking out particular restriction sites and adding regulatory sequences that can turn on inserted genes, control their action, and turn them off at the appropriate time. The genetic manipulators have also added genes for resistance to a variety of antibiotics so that bacteria that have picked up the plasmids can be separated from those that have not by adding antibiotics to the culture medium. Researchers can now call upon a wide assortment of made-to-order plasmids for cloning genes and translating them into their products in bacterial cell "factories."

Plasmids, however, are only about three to fifteen kilobases (thousands of bases) long, and they can accommodate inserted DNA fragments only up to about fifteen kilobases in length. That is a bit confining, so genetics researchers also make use of some other types of vectors. One is the *phage,* a kind of virus that infects bacteria. A phage has a circular chromosome about fifty kilobases long, and it can accommodate a foreign DNA insert of up to twenty-five kilobases. Further versatility is provided by the *cosmid,* a hybrid between portions of a plasmid and sequences from a phage that make the DNA circular and permit it to be packaged into a virus outer coat so that can infect bacteria. (The contribution of the plasmid consists of an antibiotic resistance gene and a short sequence needed to start the process of DNA copying in the host cell.) Cosmids are about twenty-five kilobases long, and foreign DNA up to fifty kilobases can be inserted into them.

Recently, geneticists from Washington University of Medicine in St. Louis have developed techniques for cloning large pieces of DNA—up to 500 kilobases or more—in plasmidlike vectors that are reproduced in yeasts. These vectors, called "yeast artificial chromosomes" or *YAC* vectors, are about fifty kilobases in size, and they will make it possible to clone DNA fragments ten times as large as those that could be cloned before.

In the cloning process, a DNA fragment with sticky ends is combined with a vector (whether plasmid, phage, cosmid, or YAC) that has been prepared by treating it with an appropriate restriction enzyme to cut open its DNA circle and leave compatible sticky ends. A DNA ligase, or joining enzyme, is added to splice together the vector and the piece of foreign

DNA, producing recombinant DNA. Then the vector, with its baggage of foreign DNA, is taken up by a host cell. The bacterium *E. coli* is used as the host in most recombinant DNA work, but other bacteria and yeasts can also be used. Just as researchers have tinkered with the vectors, they have also tailor-made the host strains, introducing safety factors to make sure that if bacteria containing recombinant DNA happened to "escape"—from a spilled culture dish, perhaps—they could not multiply and produce some exotic plague out of a science-fiction horror story. Generally the host organisms are crippled by selecting strains that require special growth conditions to survive. If one of these genetically engineered bacteria somehow got into a person's intestines, it would be unable to reproduce and would just die out.

Phages are naturally replicated by the thousands in a bacterial cell, and there are methods (referred to as *amplification*) for making a bacterium produce extra copies (up to thousands) of a plasmid. The host bacteria, in turn, can reproduce at a prodigious rate. (*E. coli* can divide every twenty minutes, which can bring the bacterial population from a single cell up to the millions overnight.) Thus, recombinant DNA methods can be used to produce substantial amounts of biochemicals and have formed the basis of a fast-growing biotechnology industry. People with genetic disorders are already reaping the benefits of this biotechnology revolution. Diabetics can take recombinant-DNA-produced human insulin, and people with pituitary disorders can grow to a normal height with the help of genetically engineered human growth hormone. Synthetic interferon is finding uses in the treatment of viral diseases and cancer and is being tested against AIDS; synthetic interleukin-2 (the natural counterpart of which is a key part of the immune system) has shown promise in the treatment of cancer. These and other biochemicals are present in natural sources in far too small amounts to be of much practical use. Recombinant DNA techniques are also being used to produce more effective vaccines, including prospective vaccines for AIDS.

In addition to the genomic library, researchers can construct another useful type of DNA library: the *cDNA library*. Starting with cells of a specific type—liver cells, for example, or brain cells—*complementary DNA* (cDNA) copies can be made of all the working messenger RNA molecules in the cell. These copies are made with the aid of reverse transcriptase, an enzyme first discovered in certain types of viruses (retroviruses) whose genetic information is carried in the form of RNA and translated backward into DNA in a host cell to make working virus genes. In contrast to a genomic library, which contains copies of all the DNA in the chromosomes (including not only genes but also regulatory sequences and a variety of apparently nonfunctional "molecular junk"), a cDNA library represents only the genetic information that is pertinent to that particular kind of cell, coding for the proteins it produces. Another distinction is that in the genomic library the DNA is chopped up randomly into pieces, and the genes that are present may be fragmented. The DNA segments in a cDNA library each contain the complete genetic instructions for a particular protein, though they are not working genes—they lack the control signals found on the original chromosomes. Inserted into a vector with the appropriate control signals and transferred to a bacterial cell, these cDNA genes can be made to work, and researchers can then isolate and study the proteins that the genetically engineered bacterial cells produce. The cDNA can also be retrieved from the cells and used as a probe to fish out natural genes.

Geneticists can also work backward from proteins, using the genetic code to figure out the possible coding sequences of messenger RNA that could direct their production and then the complementary DNA sequences. From those blueprints they can synthesize lengths of DNA—artificial genes—to be used as probes to find the location of natural genes on the chromosomes, to test the effects of variations of the base sequence, and perhaps ultimately to use in gene therapy.

Transferring Genes

Researchers have known for close to two decades that when certain kinds of viruses invade a cell, they can leave behind a portion of their DNA. The viral genes may be incorporated into the cell's own chromosomes, and under certain conditions they may be expressed, producing protein products and affecting the cell's appearance and behavior. Generally the effects of such viral genes are bad—the cell's growth goes out of control. (This is thought to be one of the mechanisms producing cancer and one of the sources of potential oncogenes.) Progress toward determining whether "good genes"—ones that did not cause cancer—could also be introduced into cells took a step forward in 1978, when Columbia University geneticists mixed ordinary cell genes with a suspension of small calcium phosphate crystals and found that mammalian cells were able to take up the genes and add them to their own genome. (This method is referred to as *DNA transfection.*)

Richard Axel, the head of the Columbia research team, notes that DNA transfection is not a very effective method for introducing genes into cells. A more successful method of DNA transfer, he has pointed out, must include four things: a delivery system; a source of selectable DNA; an appropriate recipient cell (one that lacks a gene for the product encoded by the donor DNA and is able to be transformed); and a selection system for identifying the cells that have been successfully transformed.

One way to transfer genes is simply to inject "naked genes" (pieces of DNA) into a cell. The delivery system here is not very sophisticated: a microscope, a very thin hollow needle, and a researcher with a steady hand. Surprisingly, it works. Some cells, undamaged by the rather crude injection process, take up the foreign gene, incorporate it, and even express it. A form of gene therapy has even been successfully carried out in the laboratory, using the *microinjection* technique. A gene for rat growth hormone, spliced to a mouse control sequence, was injected into fertilized mouse eggs, which were implanted into surrogate mothers. Out of 170 injected eggs, twenty-one mice were born, and seven of them carried copies of the foreign gene. The young mice grew

Procedures developed in the laboratory can lead to new products. Here a large manufacturing facility recovers and purifies many proteins made by genetically engineered microbes.

faster than usual. By adulthood, these mice were twice the size of ordinary mice, and when mated, they produced a second generation of giant mice. Subsequently, similar experiments were carried out with genes for human growth hormone and a gene for the production of a human antibody. In the case of growth hormone, although the transplanted genes were present in all of the animals' tissues, the extra growth hormone was produced in only a few organs of the "supermice," mainly the liver (the organ in which the gene normally regulated by the control sequence is chiefly expressed). The pituitary gland, which normally produces growth hormone, actually produced very little of it, probably because the body's feedback systems were sending signals that none was needed. When an antibody-producing gene was transplanted, the antibody was produced in the spleen, where such genes are normally expressed, not in the liver.

One problem with injecting genes is that so far researchers have no way of predetermining where the gene is going to be incorporated, and the specific place in a chromosome where the gene is added can have important effects on its activity. Frank Ruddle of Yale University, a pioneer in the gene injection technique, has described an experiment in which interferon genes were inserted into mouse embryos. Some of the male animals that grew from the embryos turned out to be sterile. In the tissues of these animals, the pattern of the interferon gene was different from that of the animals who had normal fertility. In some of the embryos the introduced gene had been incorporated into a spot on the chromosomes where it had interfered with the normal development of the testes.

To further complicate the situation, certain genes are capable of moving from one spot on a chromosome to another or even from one chromosome to another. Such *mobile genetic elements,* or "jumping genes," play an important role in the functioning of the immune system, bringing together parts of antibody genes that are otherwise scattered over the genome. "Jumping genes" may also play a key role in the development of an embryo, as its original cells gradually differentiate into the types that make up the various body tissues. Transfers of genes to other chromosomal locations, where they are no longer under the control of their normal genetic regulators, seems to be part of the way oncogenes are turned on, as well. Therefore, geneticists are understandably cautious about indiscriminately transferring genes, which may have unpredictable effects if they wind up at the wrong chromosomal address.

One way to ensure that a transplanted gene will be located in the right spot is to utilize the cell's own mechanisms for repairing DNA. If the replacement gene is designed to closely match the sequence of bases in the cell's own DNA, it will home in on the matching segment. The use of suitable enzymes to make cuts in the transplanted gene turns on the cell's natural repair mechanisms, which will then cut out the faulty gene and splice in the new one. A similar method can also be used to inactivate particular genes.

Viruses may provide a more efficient delivery system for gene transfer than simple injection or transfection. A virus that normally infects mammalian cells comes already equipped with the means of penetrating into the cells and getting its genetic information replicated by the host. Experiments on animals have shown that viral genetic material is introduced quite efficiently into embryos: 30 to 50 percent of the infected embryos are found to incorporate the viral DNA into their genome. The viral genes may be activated during differentiation; whether they are or not seems to depend on the place on the chromosome where they are located. Genes could also be carried by viruses into particular types of body cells in children or adults. Some viruses tend to home in on specific organs; or scientists could attach chemical tags that would match up with cell surface receptors and help the viruses ferrying in the genes to "dock" at the appropriate cell.

An important consideration in using viruses to transfer genes is to be sure the viruses themselves do not produce any harmful effects. The last thing genetic engineers want to do is to give a patient a dangerous infection or cause cancer in the act of correcting a genetic disorder. Thorough animal experiments on potential virus vectors to establish that they do not cause disease will be needed before such gene transfers can be tried on humans. Researchers are also investigating the possibility of inserting a "suicide gene" into a virus vector. This gene would code for an enzyme that could convert a harmless substance into a poison inside the host cell. Then, if the virus vector produced a cancerous transformation in a cell, the patient could be given a drug to activate the "suicide gene" and wipe out the affected cells.

Still another possible way of transferring genes is *cell fusion.* Back in the 1960s it was discovered that viruses can cause two cells to fuse and mingle their contents, producing a single hybrid cell with a double set of chromosomes. Today a strain called Sendai virus is routinely used to produce cell fusion. (Treatment with ultraviolet inactivates the virus's ability to infect cells, but it is still able to make them fuse.) Some exotic hybrid cells have been produced, including mouse-rat, mouse-hamster, mouse-chicken, and mouse-human hybrids. When hybrid cells are grown in a culture dish, they tend to lose chromosomes with each successive division, rapidly at first but then more slowly, until finally stable cell lines are obtained. Such hybrid cells retain some of the chromosomes from each original parent cell, and tests show that all the chromosomes are working.

Hybrid cells produced by cell fusion are excellent tools for mapping chromosomes. For example, researchers can test a hybrid cell culture for the presence of a particular protein—say, a human enzyme. If the protein is found, an examination of the karyotype will reveal which human chromosomes are present and which ones have been lost by the hybrids. Other hybrid cell clones will have different assortments of chromosomes. By testing and comparing them, it is possible to determine by processes of elimination which chromosome contains the gene that codes for the protein tested. Hybrid cells are also yielding information on how genes are turned on and off as an organism develops. Researchers at the National Cancer Institute, for example, have found that when mouse blood cells and human blood cells are fused, the hybrid produces both mouse and human hemoglobins. But if a mouse blood cell is fused with a human skin cell, only mouse hemoglobin is formed. Although the human skin cells still retain the hemoglobin gene, it is apparently turned off, and the mouse blood cell environment does not turn it back on.

For gene therapy, some of the patient's cells (from an organ that would normally produce the lacking gene product) could be fused with cells containing working genes for the needed product. These corrective cells could be normal human cells or (to minimize rejection reactions by the patient's body) some of the patient's own cells, into which the appropriate genes have been inserted. The fused cells would be grown in culture, and the hybrids with the right activity could be located by testing for the product. These hybrids could be futher multiplied in culture and injected into the patient.

8

Gene Therapy

David spent nearly all of his short life separated from the world by sheets of plastic. His parents had had two children before him: a healthy girl and then a boy who died when he was only six months old, of a seemingly minor infection. Doctors diagnosed the dead boy's problem as severe combined immune deficiency (SCID), of an X-linked type. Unless the disorder was a new mutation, any other male children of the family had a one in two chance of inheriting SCID as well. The parents wanted another child and decided to take the chance. When amniocentesis revealed that their new fetus was a boy, the doctors prepared for heroic measures. The baby was delivered by Caesarean section in a specially sterilized hospital "clean room" and immediately transferred to a plastic isolator. The tent-like chamber was stocked with sterilized formula, diapers, and other baby supplies; gloves built into portholes in the walls allowed people to pick up and handle baby David. At the time, no one was sure how long he would have to stay in his plastic bubble. Tests might show that he was not suffering from SCID after all. (That was what had happened with the first germ-free baby, who had been born in a similar sterilized hospital environment in England, not long before.) If David did have an immune problem, he might recover spontaneously and begin to produce working white blood cells on his own; that had also happened to other children in the past. And if not, there was an experimental new technique of bone marrow transplants that might be used to give David functioning immune defenses. All in all, it was estimated that the child would have to remain in his plastic isolation tent for about two years at most.

It didn't work out that way. The tests quickly showed that David had no functioning T cells and only a few B cells—a bad case of SCID. As the months went by, he showed no signs of producing any immune defenses on his own. Family members were tested for tissue antigens, but unfortunately none of them was a good enough match to David's tissues to try a marrow transplant.

David lived on in his plastic tent. His little isolator was replaced by a larger one, and he was sent home with instructions to his parents on how to care for him while keeping him safe from contact with the usual germ-filled air, water, and food. The doctors were afraid that the boy would not develop normally, growing up without ever feeling the warmth of a human body or the touch of bare skin, without being able to run and play outdoors like other children. But David grew into a cheerful, affectionate child, bright for his age but also capable of throwing an occasional childish tantrum. Additions to his isolator permitted him to jump and climb; he quickly learned to take refuge out of reach of the built-in gloves when a mischievous prank had earned him a spanking. He started school on schedule, with a telephone link to the classroom. When David was six, NASA built him a miniature spacesuit that made it possible for him to explore his family's house for the first time and even to go outside with a portable air-filtering device trailing along behind.

Gradually, in spite of all the precautions, a few germs slipped in. By the time David was seven, several dozen kinds of bacteria were sharing his plastic bubble. They were harmless varieties and didn't seem to bother him (he had never been sick), but he still had no defenses against disease germs. More years went by. People all over the world knew about "David the bubble boy" and watched for the birthday pictures and articles about him. Gradually, though, David grew moody, brooding about his isolation from everyday life. By the time he was twelve, doctors were working on a new marrow transplant technique. Even mismatched marrow could be used, by treating it to remove the T cells responsible for the potentially deadly graft-versus-host reaction. When the possibilities and risks were explained to David and his parents, they agreed to go ahead, using bone marrow from his healthy older sister.

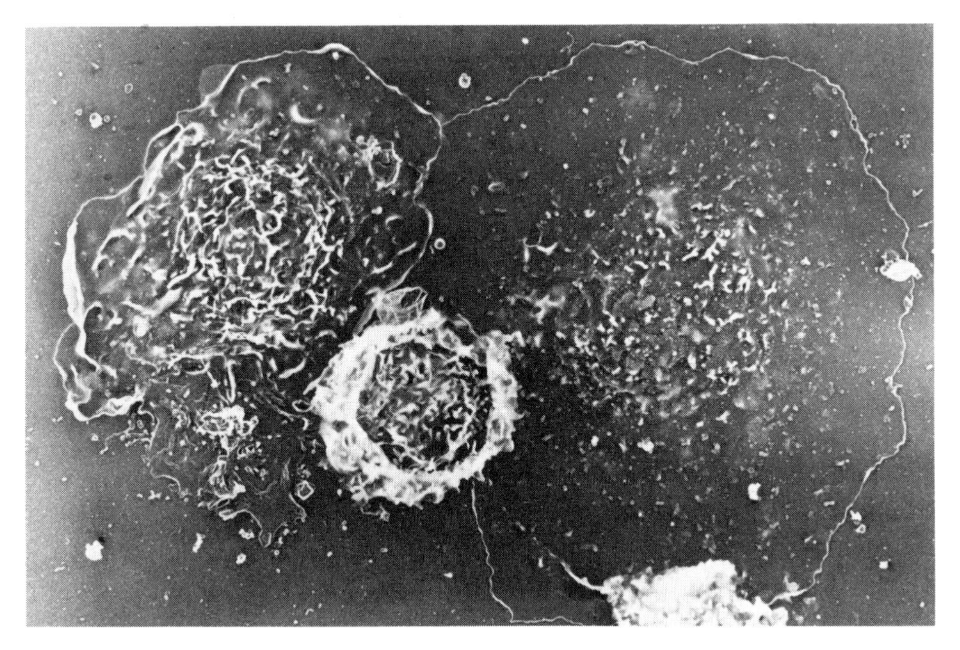

The macrophage, a type of white blood cell that can engulf invading bacteria, is a key part of the human immune system.

After the operation the doctors were hopeful at first; there seemed to be some signs that David was developing immune defenses. But then he became ill—so ill that it was finally decided to take him out of the bubble to be able to treat him more effectively. Soon after he was finally able to hug and kiss his family without a sheet of plastic between them, David died. No one had realized it at the time, but his sister's marrow cells contained a virus, of a type that can cause cancer. Without working T cells, David had had no defense against the tumors that sprouted all through his body.

A form of SCID, *ADA deficiency,* is currently the best prospect for treatment by gene therapy. Unlike the X-linked disorder that affected David, ADA deficiency is an autosomal recessive disorder, and it is even rarer than the sex-linked forms: Only eight to ten children in North America currently have it. They lack the enzyme adenosine deaminase; as a result, substances poisonous to T and B lymphocytes build up in their blood. The toxic substances kill off these important white blood cells, leaving the sufferers without defenses against disease-causing microbes and cancer. When a child with ADA has a sibling whose tissues are a good match, bone marrow transplants have a better than 80 percent chance of curing the disorder. But only about one in four children with ADA has a healthy sibling (or other relative) who is a good enough match. Of the rest, 90 percent die within their first fifteen months.

ADA deficiency is a good candidate for gene therapy for several reasons. First of all, it is a single-gene disorder, and the gene to correct it is available; it has been isolated and cloned. Bone marrow stem cells (the generalized cells that develop into various kinds of blood cells) can be isolated readily from bone marrow. If they are treated to introduce the normal gene, then injected into the patient's bloodstream, they will travel to the bone marrow and settle down, differentiating into the T and B cells the body lacks. Even if the gene transfer is successful in only a small percentage of the stem cells, that should be enough; the normal cells have a survival advantage since they have the enzyme to protect them from being poisoned.

Geneticists are still working out the details of gene transfer procedures in animal experiments. They have successfully transferred the ADA gene to human cells in culture, and these cells expressed the gene, producing the enzyme. Researchers encountered problems at first when they tried to go from cell cultures to living animals; often the genes remained inactive. But in recent experiments on monkeys conducted by a team of National Heart, Lung, and Blood Institute researchers headed by W. French Anderson, the transferred human gene directed the production of ADA. The effect lasted only for about 120 days, the lifetime of the transformed bone marrow cells, and it still must be demonstrated that the "disarmed" retrovirus used to carry the ADA gene into the bone marrow cells does not cause any harm. But in mid-1987, Anderson's team presented a federal government review board with a preliminary draft of a proposal to try the procedure on human SCID patients.

Meanwhile, other research groups have been pursuing different strategies. A group at the University of California, San Diego, is trying to develop a technique for introducing genes into liver cells instead of bone marrow. At Scripps Clinic and Research Foundation in La Jolla, California, researchers have used gene therapy to correct the defect typical of Gaucher's disease in cultures of *fibroblasts,* a type of connective tissue. Researchers at Harvard University are also working with fibroblasts, using a chemical technique instead of a viral vector to transfer the gene for human growth hormone in cell cultures. The transformed cells were then injected into experimental mice at various sites. Those injected into the abdominal cavity seem to work best, producing growth hormone for at least two weeks. The Harvard geneticists call their technique *transkaryotic implantation.*

Any attempts to apply gene therapy to humans must first be approved by local review committees and the federal government, according to strict guidelines issued in 1985. They were drawn up after some unsupervised attempts were made when scientists had very little awareness of all the risks and complexities involved.

The first gene transfer experiment started by accident. Close to fifty years ago, it was discovered that warts on the skin of cottontail rabbits in western Kansas are caused by a microbe called Shope papilloma virus. This virus does not seem to produce any disease in humans (not even warts). In fact, its discoverer, Richard Shope, was so convinced of its safety that hc injected himself with it in 1933; he reported that nothing happened.

In 1959, biochemist Stanfield Rogers discovered that something does happen to people exposed to Shope papilloma virus. He demonstrated that the virus causes the production of arginase, an enzyme that breaks down the amino acid arginine. In rabbits infected with the virus, the warts produce large amounts of arginase, and the rabbits' blood contains unusually low amounts of arginine; the viral enzyme is working in their bodies.

Rogers began a follow-up study of about seventy-five laboratory workers who had been exposed to the Shope papilloma virus during preceding decades. *All* of them proved to have unusually low levels of arginine in their blood. Apparently they had all picked up the virus gene. Yet they were healthy. Neither the virus nor the low arginine levels seemed to be causing any problems. Rogers concluded that he had discovered "a form of treatment with no known disease."

Eventually a disease turned up. In 1969, geneticist Joshua Lederberg of Stanford University sent Rogers a note calling his attention to two young sisters in Germany with a genetic disorder involving a lack of arginase. Their blood serum levels were ten times higher than normal, and they were severely mentally retarded. Rogers wrote to the German doctors who were treating the case, offering to "challenge" the new disease with Shope papilloma virus. Small skin samples from both girls were flown to Rogers' laboratory in Oak Ridge, where they were cultured and infected with purified arginase. Sure enough, the cultured cells began to produce arginase. Meanwhile, however, the girls were getting worse. Rogers sent some samples of the virus to Germany, and the doctors there tried to treat the girls with virus injections. Their serum arginine levels were somewhat lowered, but the girls remained mentally retarded. Apparently it was too late to repair the brain damage.

Then the parents decided to ignore the advice of the genetic counselors and had another baby girl in 1971. She, too, was soon diagnosed as suffering from arginemia. A single dose of the virus was given to the baby when she was a few months old, but she did not improve. A hail of medical criticism rained down on Rogers for using an untested treatment on human beings. Rogers argued that forty years of experimentation had already shown the virus was safe, and since there was no other treatment for the disease, it was worth the chance—the sisters were doomed otherwise. Eventually it was discovered that the treatment of the third child had not been a real test. The viruses in the dose she received had died in the mail on the way to Germany, and she never received any live virus at all.

The next attempt at gene therapy was more sophisticated but equally unsuccessful—and unfortunate. Martin Cline, a specialist in blood disorders at the University of California, Los Angeles, did his preliminary experiments with mice. First he set out to use gene transfer techniques to make mice resistant to the anticancer drug methotrexate. This drug kills cancer cells by preventing them from utilizing a vitamin, folic acid, but it is also rather toxic to normal cells and can cause severe destruction of a cancer patient's bone marrow. If the normal marrow cells could be made resistant to methotrexate, then larger doses of the drug could be given to wipe out the cancer cells more effectively. It was known that some cells are naturally resistant to methotrexate because they have multiple copies of a key enzyme of folic acid metabolism, dihydrofolate reductase (DHFR). Cline exposed mouse marrow cells to DHFR genes, then injected the treated marrow cells back into the mice. He subjected the mice to daily injections of methotrexate, so that the ordinary marrow cells would be killed off and those that had taken up extra DHFR genes would survive and multiply, eventually taking over the marrow cavity. The mice did become methotrexate-resistant, and their marrow cells were found to have extra DHFR genes.

Other researchers were unable to repeat Cline's experiments, and they objected that he had not demonstrated that gene transfer had really occurred. Perhaps some of the original cells had mutated and begun producing extra DHFR genes on their own. After another series of experiments, in which Cline transferred a different gene to mouse cells (also criticized by some of Cline's fellow scientists), the UCLA researcher submitted an application to his university review board to transfer normal hemoglobin genes to patients with sickle-cell anemia. Meanwhile, he had similar applications pending at hospitals in Israel and Italy for the treatment of thalassemia, another hereditary hemoglobin disorder.

In mid-1980, approval from the foreign hospitals came through. Cline treated bone marrow cells from a young woman in Jerusalem with beta-hemoglobin genes, together with viral genes that Cline thought would give the transformed marrow cells a survival edge over the untransformed cells. While the cell cultures were incubating, the woman's lower legs were treated with radiation to kill off some of her marrow cells and make room for the reinjected cells, and then Cline injected about 500 million treated marrow cells, perhaps 5000 to 50,000 of which might have been expected to have picked up the new genes. Another injection was given on the following day. A few days later, Cline administered a similar treatment to a young thalassemia patient in Naples. Meanwhile, back at UCLA, the committee rejected Cline's application to treat sickle-cell patients, saying that more animal studies were needed. Word about the gene transfer treatments leaked out, and many scientists were appalled—including the molecular biologist who had supplied the beta-hemoglobin genes, not realizing they were going to be used on humans. Some doctors argued that with a life-threatening disease for which there was no other treatment option, experimenting with gene transfer seemed reasonable. As for possible dangers—that a gene might go out of control and cause cancer—these possibilities seemed hardly more fearful than the girls' already very limited life expectancy.

Cline's bold experiments did not work. There was no sign that either of the thalassemia patients produced normal beta-hemoglobin after the treatment. Cline himself suffered serious repercussions. The National Institutes of Health found him guilty of violating federal guidelines for human experimentation and for recombinant DNA research because he had slightly modified his procedure at the last minute and had not gone back to the hospital committees to have the change approved. The funding for Cline's research was cut back drastically, and his further research projects were subjected to hampering red tape.

The 1985 federal guidelines clarified the procedures to be followed for any attempts at gene transfer in humans, but some of the bioethical issues raised are still far from resolved in the minds of many, both scientists and lay people. How do we determine at what point there has been enough animal study to proceed to work with humans? Are long-term risks really important when a genetic disorder is dooming a patient to death—or reducing the quality of life so much that it scarcely seems worth living?

In retrospect, the fumbling early attempts do indeed seem premature; a number of technical difficulties still remain to be worked out. It seems likely that improvements in technique will ultimately overcome the low rates of gene transfer and the inactivity of some transferred genes in their new cell homes. Early in 1988, a British research team reported the successful transfer to adult mice of the human gene that directs beta-hemoglobin production. Not only were the transplanted genes functioning successfully in some of the mice nine months later (half a mouse's normal life span), but they were functioning only in the appropriate kinds of cells—red blood cells and the bone marrow cells that produce them. As for the ADA gene, which often remains inactive after transfer to mouse cells, W. French Anderson notes that it was activated in monkeys because they are so much closer genetically to humans and believes that activation should present even less of a problem with human patients. Indeed, for many genetic disorders, full effectiveness may not be necessary. In autosomal recessive disorders due to the lack of an enzyme, carriers usually are healthy even though they produce only half the normal amount of the enzyme.

Bone marrow cells have thus far been the main target of gene therapy experiments because they can be fairly easily removed from the body and later replaced; they also have the potential to multiply rapidly and produce large quantities of their genetically engineered product. Recent experiments at Massachusetts Institute of Technology and Harvard University point to another convenient target for future gene therapists: skin cells. These cells—like those from bone marrow—have the potential for rapid reproduction, and they can be grown in a tissue culture and then grafted back onto the body surface. The Massachusetts researchers have transferred the gene for human growth hormone to human skin cells. Not only did these cells produce the hormone in culture, but they continued to manufacture it when they were grafted onto animals. Patches of cultured skin, with suitably modified genes, could thus be used to provide a supply of an enzyme or hormone to people whose bodies cannot produce it.

Genetic disorders that affect the brain present special problems and challenges to the gene therapist. The replacement genes to correct hereditary blood disorders may settle naturally in the bone marrow and other blood-producing tissues, and hormone deficiencies may be cured by transferred genes incorporated into liver or even skin cells, far from the organs that normally produce them. But gene therapy of nervous system disorders may require delivering the replacement genes to the nerve cells themselves. This has already become evident with bone marrow transplants for Lesch-Nyhan patients. Their kidney problems are helped, but not the terrible mental damage.

Researchers at the University of Michigan have suggested that instead of retroviruses, herpes simplex type 1 viruses can be used as vectors to carry replacement genes into the body. This is the virus that causes cold sores; people tend to get recurring herpes attacks because the virus can migrate to nerve cells and enter them, remaining dormant until it is activated again. In preliminary experiments on rats, the Michigan team cloned the human gene for HPRT (the missing enzyme in Lesch-Nyhan syndrome) in herpes simplex type 1 viruses, then exposed rat nerve cells to the virus. The infected cells began to produce HPRT. These experiments are promising but still very preliminary. For one thing, since the herpes virus can cause disease, it will have to be extensively modified to make it harmless before it can be used on humans.

Another sobering concern is that gene therapy treatments must begin soon enough, before irreversible damage is done. What good would it do to prolong the life of a child with Tay-Sachs disease, for example, if all the mental faculties had already been destroyed by the poisons that accumulated before the genetic defect was repaired? In some cases the damage begins even before birth, at such an early stage of development that the mother might not even be aware that she is pregnant. The need to intervene before key organs are damaged beyond repair may be a limitation inherent in the approach of treating genetic disorders by altering somatic (body) cells.

Some people have raised another bioethical question about such treatments: If we can help people with genetic disorders to live a normal life by altering the genes of their somatic cells, they will live to have children and will probably pass the disorder on to many of them. Will this increase the frequency of genetic disorders and create problems for future generations? Yet what is the alternative? Should we not treat them, dooming them to suffer and die? Should we treat them but forbid them to reproduce? Have we the right to do that? Surely people with genetic disorders have not forfeited their own rights just because they have been unlucky in the genetic lottery. For that matter, who knows what harmful genes outwardly normal people may be carrying recessively?

One alternative might be to treat not only somatic cells but germ cells as well, so that the genetic corrections could be passed on to future generations. Already researchers are making progress on the animal level. In 1987, geneticists at California Institute of Technology and Harvard University announced that they had used gene therapy to cure mice of a fatal hereditary neurological disease. Mice with the *shiverer* mutation fail to produce a protein normally found in the insulating sheaths of nerve cells in the brain and spinal cord. Within three months after birth, they die from uncontrollable shaking. The Caltech team isolated the gene that directs the formation of the missing protein, MBP, attached special DNA sequences to regulate the gene's expression, and cloned it. Then they inserted 200 copies of the gene into each of 320 fertilized eggs from mice carrying the shiverer mutation and implanted them into surrogate mothers. Only two mice were born; the others died in the womb. But one of the gene-treated mice showed only mild shivering symptoms; she was found to be producing some MBP, though not as much as the normal level. In matings she passed the cloned gene on to her offspring.

These experiments are encouraging, but they also hold some warnings. With the present techniques, the rate of success in germ cell gene transfer is very low (only one success out of more than three hundred tries), and the researchers found that the gene insertion is apparently random: One of the embryos that died was unable to make collagen, an important protein in connective tissues, because its cells had incorporated the transferred gene into the middle of the DNA sequence coding for collagen. Considerable work lies ahead before researchers will be ready to try such therapies on humans. Indeed, federal guidelines for gene therapy experiments currently do not permit this type of experiment. Only somatic cells can be treated, not germ cells or embryos. But as the knowledge and skills of genetic engineers increase, those guidelines may be changed.

Some of the exciting current experiments in gene therapy are focused on fighting infectious diseases by genetically increasing the body's resistance to the microbes that cause them. This sort of tinkering is not really new. Plant and animal breeders have long been using various breeding techniques to make their stock hardy and disease resistant. What is new is the idea of directly changing and manipulating the genes themselves, rather than taking advantage of naturally arising mutations. Implanted genes have been used to give plants protection against a number of viral diseases. Now Enzo, a biotechnology firm, is working on an intriguing gene therapy treatment for AIDS. The technique involves producing "antisense genes"—sequences of RNA complementary to the messenger RNA of the AIDS virus. Enzo plans to introduce these genes into bone marrow cells from people with AIDS, then reintroduce the treated cells into the body. These

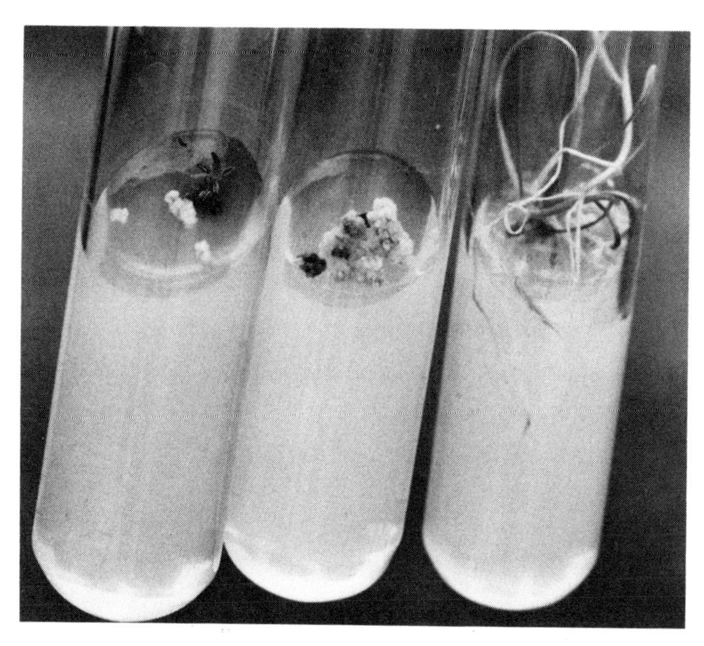

In these test tubes with culture medium, wheat plants are being grown from cells with genetically engineered chromosome changes.

cells would be protected from attack by the AIDS virus because the antisense sequences would bind to the viral mRNA, effectively preventing it from producing its proteins. The protected marrow cells would multiply and produce T cells that would resist the attack of the virus and thus protect the infected person from its harmful effects.

If this approach proves successful, researchers will have taken gene therapy a step further: Rather than correcting a hereditary defect, they will have improved on the natural design, making the body better fitted to survive under the attack of the AIDS virus. This kind of improvement has already been going on in nature for millions of years, through the process of evolution by natural selection acting on mutations that arose randomly. But natural evolution has been a slow and chancy process. Now we are proposing to speed up the changes enormously and to direct their course. The implications are awesome. We may be gaining the knowledge to alter ourselves, but will we use it wisely?

At this point, the public seems to have mixed feelings about the bioethics of gene therapy. According to a survey of American attitudes toward genetic engineering reported by the Office of Technology Assessment in 1987, while Americans still favor strict regulations, they also strongly support continued research into genetic engineering. The conflicting feelings show up most sharply when abstract generalizations are contrasted with specific applications. When asked if they felt that changing the genetic makeup of human cells was morally wrong, for example, 42 percent of those surveyed said they thought it was. But then, when the same people were asked if they would be willing to have their child undergo that type of therapy to cure a usually fatal disease, 86 percent said yes. The percentage of people who favored continued research into genetic engineering (82 percent) cut across party lines, age groups, educational levels, and economic groups.

9

Engineering Our Future Evolution

It sounded like science fiction. Late in 1986, gene splicers at the University of California, San Diego, announced that they had successfully transferred a firefly gene to tobacco cells and had grown plants that glowed in the dark. You might wonder why anyone would want to produce glowing tobacco plants. (To help moonlighting farm workers harvest the leaves, perhaps?) Actually, the researchers chose tobacco because the plants are easy to grow in the laboratory, a lot is known about their genetics, and whole plants can be grown from single cells. A few months later, the same research team described experiments in which the firefly gene for luciferase, the enzyme that produces the glow, was transferred to cultured monkey cells. Far from being just a scientific curiosity, the new firefly gene hybrids promise to be a valuable research tool for developing genetically engineered vaccines, hormones, blood-clotting agents, and monoclonal antibodies. They may provide markers for identifying and replacing defective genes in patients with genetic disorders.

The unusual experiments grew out of twenty years of research by UCSD biochemist Marlene DeLuca on how the firefly enzyme produces light. The work of collecting and grinding up fireflies was so expensive and laborious that she combined her efforts with molecular geneticist Donald Helinski to isolate and clone the firefly gene. They discovered the glow enzyme was coded by a single gene that could be easily manipulated; that finding suggested still further ideas. Splicing the luciferase gene to the regulatory switch of a gene from a virus that infects plants, the research team then inserted the gene with its new controls into a plasmid from the *Agrobacterium,* a microbe that produces tumors in plants. When this plasmid was incubated with tobacco leaf cells, some of the cells picked up the recombinant plasmid. The transformed cells were then grown into plants. Watering the plants with a solution containing another firefly substance, luciferin (the normal substrate of luciferase), produced a dim

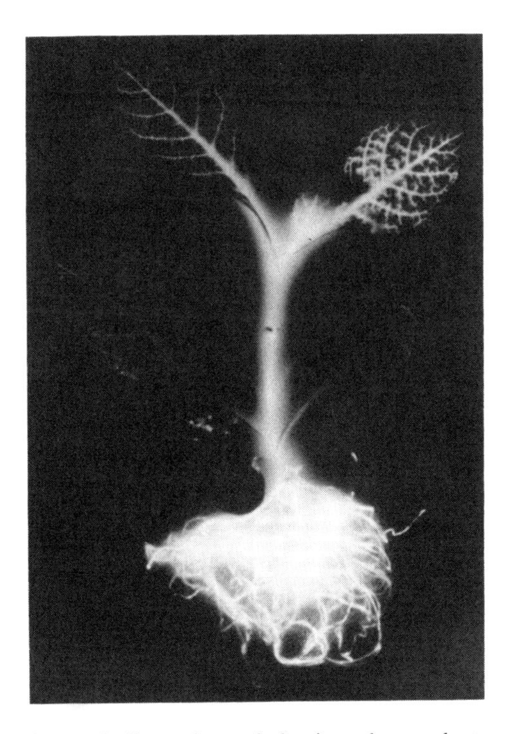

A genetically engineered glowing tobacco plant.

glow that could readily be picked up by light detectors or photographic film and could be seen by people whose eyes had adapted to the dark. A similar technique was used to transfer the gene to monkey cells, which could also be made to glow by adding luciferin.

The potential of the new hybrids is exciting. Researchers could introduce the gene into the DNA of a single cell and then follow its daughter cells through many generations as they divide and differentiate, producing tissues. By joining the firefly gene to specific plant or animal genes, researchers could study the conditions in which these genes are turned on simply by watching what parts of the organism light up and when.

The real possibilities of this achievement are thrilling enough, but the idea of transferring firefly genes to plants or animals is one that tickles the imagination and sets one dreaming about all sorts of intriguing possibilities. If scientists can already combine the genes of two such different creatures as a firefly and a tobacco plant, what else can they do? Will it be possible some day to change our own heredity, perhaps even to improve the human species?

Would you like to be able to synthesize your own food, like a plant? That idea probably is destined to remain only science fiction. Very likely, the appropriate genes could be incorporated for humans to store chlorophyll in their skin and capture some of the energy of sunlight. But the process would not generate nearly enough energy to meet the needs of an active human being, who moves around and maintains an even body temperature. Plants can get along all right because they do not move around and have more modest energy requirements than we do.

How about being able to breathe under water? That idea may be more feasible. We actually have the genes for gills, somewhat like those of fish. Gill-like structures appear briefly during the development of every human embryo, but then the genes are turned off, and the gills disappear. One day we may learn how to turn inactive genes back on.

If we do, a very valuable addition to human capabilities would be *regeneration,* the ability to regrow parts that have been damaged or lost. We have that ability at an early stage of development, and some of our cells retain it throughout life—liver cells and skin cells, for example. But most of our body tissues and organs cannot be replaced—now.

A mastery of the genetic on-off switches and other controls might also give us the keys to the aging process and to stopping or even reversing it. We might even achieve a virtual immortality.

Would these be improvements? That is not as easy to decide as it might seem. Take intelligence, for example. Today's scientists cannot even fully define it, much less determine which aspects of intelligence are the most valuable, and they are furiously debating the ways in which intelligence is currently measured. If we could increase intelligence (assuming we could figure out what it is and what we could do to modify the many interacting genes and environmental factors that contribute to it), would that be a good thing or not? Would intelligence bring wisdom? And what might be lost in the genetic juggling?

For that matter, some geneticists speculate that even in using gene therapy to repair defective genes, we may produce unanticipated negative effects. Engineering out the sickle-cell gene, for example, would save the hapless homozygotes a great deal of misery, but the heterozygous carriers would lose their advantage in areas where malaria is a major health problem. Who knows what other advantages apparently negative traits may carry?

Our past experiments in genetic tinkering—using techniques of animal and plant breeding before researchers developed the ability to splice genes—have not been uniformly successful. And some of our most brilliant successes have proven to have unexpected drawbacks. Plant breeders have developed new crop strains that were much hardier than natural varieties and gave vastly increased yields. Farmers enthusiastically adopted them—and then saw their whole crop wiped out a few years later by a new fungus disease or a shift in weather patterns. The new plants were too specialized and lost their survival advantage when conditions changed. When we gain the ability to change our own genome at will, we may need to curb our enthusiasm. If everyone jumps on the bandwagon and incorporates an attractive new modification, we may rob the human genetic stock of the natural variability our species may some day need to survive.

Developing a new crop strain in the laboratory: the vial contains peach shoots grown from cells in culture medium.

It could be argued that if environmental conditions change and we need new combinations of genes, we can use genetic engineering techniques to produce them. But too great a dependence on high-tech solutions may leave us dangerously vulnerable to some future catastrophe. Fortunately, it is unlikely that enough people will be able to agree on what changes are desirable to seriously reduce our genetic variability.

Who is to decide just what changes in the human genome are really improvements and which ones should be incorporated, if we gain the knowledge and skills to do so? Scientists sometimes tend to get carried away with enthusiasm for their work and are not always the most objective judges of what is prudent. But politicians do not have a very good track record for making ethical decisions, either. And the idea of making genetic improvements brings uncomfortable memories of the "eugenic" experiments in Nazi Germany, where the "final solution" was applied to millions of people judged to be inferior, and a misguided Lebensborn project attempted to breed perfect Aryans. It is not surprising that the average person views the potentials of genetic engineering with hope and also with some gut-level misgivings.

In a 1985 Harris poll sponsored by *Business Week,* about two-thirds of those questioned answered yes when asked if treatments altering genes to cure people with fatal diseases should be allowed to go ahead. Nearly as many agreed that if they were found to be carriers of a genetic disease and could have their genes altered to protect their children and future generations, they would do so. But the idea of altering genes to improve one's children—to make them smarter, physically stronger, or better-looking—made them pause. A resounding 88 percent felt that would be going too far.

FURTHER READING

BOOKS

Anderson, Bruce L. *The Price of a Perfect Baby*. Minneapolis: Bethany House, 1984.

Anderson, L. Kerby. *Genetic Engineering*. Grand Rapids: Zondervan, 1982.

Bains, William. *Genetic Engineering for Almost Everybody*.London: Penguin Books, 1987.

Baskin, Yvonne. *The Gene Doctors: Medical Genetics at the Frontier. New York: William Morrow, 1984.*

Carney, Thomas P. *Instant Evolution: We'd Better Get Good at It. Notre Dame: University of Notre Dame Press, 1980.*

Davis, Bernard. *Storm Over Biology*. Buffalo: Prometheus, 1986.

Fletcher, John C. *Coping with Genetic Disorders*. San Francisco: Harper & Row, 1982.

Friedmann, Theodore. *Gene Therapy: Fact and Fiction*. Anbury Public Information Report. Cold Spring Harbor Laboratory, 1983.

Grobstein, Clifford. *Science and the Unborn: Choosing Human Futures. New York: Basic Books, 1989.*

McCuen, Gary E. *Manipulating Life: Debating the Genetic evolution. Hudson, Wisconsin: Gary E. McCuen Publications, 1985.*

Nichols, Eve K. *Human Gene Therapy*. Cambridge, Massachusetts: Harvard University Press, 1988.

Nossal, G. J. V. *Reshaping Life*. Cambridge: Cambridge University Press, 1985.

Office of Technology Assessment, Congress of the United States. *Mapping Our Genes*. Washington, D.C.: U.S. Government Printing Office, 1988.

President's Commission for the Study of Ethical Problems in Medicine and Biomedical and Behavioral Research. *Screening and Counseling for Genetic Conditions. Washington, D.C.: U.S. Government Printing Office, 1983.*

Silverstein, Alvin and Virginia. *AIDS: Deadly Threat*. Hillside, New Jersey: Enslow Publishers, 1986.

Thompson, James S. and Margaret W. Thompson. *Genetics in Medicine,* 4th ed. Philadelphia: W.B. Saunders Company, 1986.

Weiss, Ann E. *Bioethics: Dilemmas in Modern Medicine*.Hillside, New Jersey: Enslow Publishers, 1985.

PERIODICALS

Adler, Jerry. "Every Parent's Nightmare." *Newsweek*, March 16, 1987, pp. 57-66.

Allen, Charlotte Low. "Feats to Concoct the Flawless Body." *Insight*, July 11, 1988, pp. 8-11.

Asher, Jules. "Born to Be Shy?" *Encyclopedia Science Supplement 1989. Danbury, Connecticut: Grolier, 1988.*

Brower, Montgomery and Meg Grant. "A Desperate Quest into a Family's Past." *People Weekly*, January 11, 1988, pp. 28-33.

Dworetzky, Tom. "Opening New Frontiers in Molecular Biology. "*Discover*, March, 1987, pp. 14-15.

Elmer-DeWitt, Philip. "The Perils of Treading in Heredity." *Time*, March 20, 1989, pp. 70-71.

Hamilton, Joan O'C. and Reginald Rhein, Jr. "The Gene Doctors." *Business Week*, November 18, 1985, pp. 76-85.

Holzman, David. "Gene Therapy Poised for Takeoff." *Insight*, November 7, 1988, pp. 54-55.

McAuliffe, Kathleen. "Predicting Diseases." *US News & World Report, May 25, 1987, pp. 646-670.*

Murphy, Jamie. "Conquering Inherited Enemies." *Time*, October 21, 1985, pp. 59-60.

Otten, Alan L. "Price of Progress: Efforts to Predict Genetic Ills Pose Medical Dilemmas." *Wall Street Journal*, September 14, 1987, p. 2-1.

Squires, Sally. "Righting Nature's Wrongs. Gene Therapy: On theThreshold of Medicine's Next Revolution." *The Washington Post,* October 30, 1985, pp. 10-13.

Thompson, Dick. "Coming: A Historic Experiment." *Time*, February 13, 1989, p. 64.

Virshup, Amy. "Perfect People: The Promise and Peril of GeneticTesting." *New York Magazine*, July 27, 1987, pp. 26-34.

Weaver, Robert F. "Beyond Supermouse: Changing Life's Genetic-Blueprint." *National Geographic*, December, 1984, pp. 818-847.

Weiss, Rick. "Predisposition and Prejudice." *Science News*, January 21, 1989, pp. 40-42.

Addresses

Helpful information can be obtained from:

American Diabetes Association
505 Eighth Avenue
New York, NY 10018

American Heart Association
7320 Greenville Avenue
Dallas, TX 75231

American Lung Association
1740 Broadway
New York, NY 10019

Cystic Fibrosis Foundation
6931 Arlington Road
Bethesda, MD 20814

March of Dimes
Birth Defects Foundation
1275 Mamaroneck Avenue
White Plains, NY 10605

Muscular Dystrophy Association
810 Seventh Avenue
New York, NY 10019

National Down Syndrome Society
141 Fifth Avenue
New York, NY 10010

Planned Parenthood Federation of America
810 Seventh Avenue
New York, NY 10019

Index